AF328545

Secrecy and Safety

DISABILITY HISTORY OF SCIENCE, MEDICINE,
AND TECHNOLOGY

Mara Mills, Wayne Tan, and Jaipreet Virdi, Series Editors

Secrecy and Safety

A Cultural History of Seizures

in Modern America

Rachel Elder

JOHNS HOPKINS UNIVERSITY PRESS   BALTIMORE

Johns Hopkins University Press
2715 North Charles Street
Baltimore, Maryland 21218
www.press.jhu.edu

Library of Congress Cataloging-in-Publication Data is available.

A catalog record for this book is available from the British Library.

ISBN 978-1-4214-5294-4 (hardcover)
ISBN 978-1-4214-5295-1 (ebook)

*Special discounts are available for bulk purchases of this book. For more
information, please contact Special Sales at specialsales@jh.edu.*

EU GPSR Authorized Representative
LOGOS EUROPE, 9 rue Nicolas Poussin, 17000, La Rochelle, France
Email: Contact@logoseurope.eu

Contents

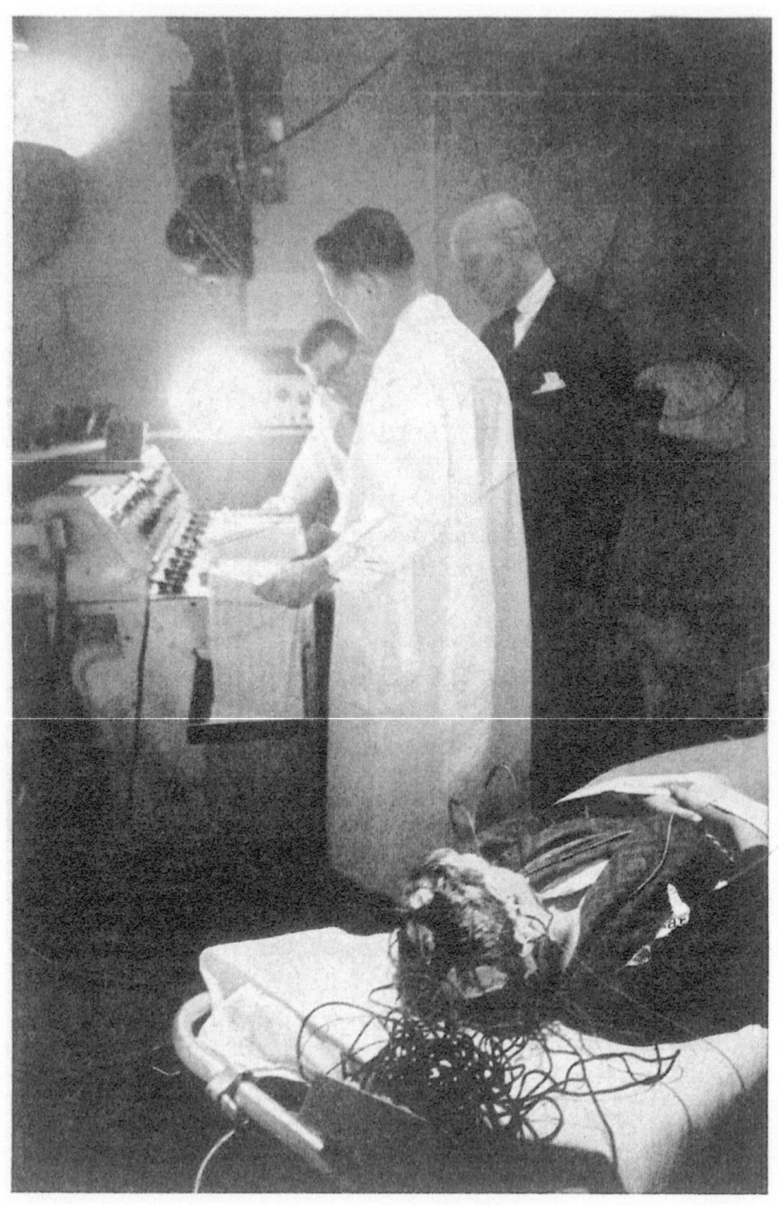

Patient undergoing electroencephalographic reading, Montreal Neurological Institute, ca. 1956. Courtesy of *Maclean's* magazine

Preface

Most of us have a personal connection to the projects we pursue. My own relationship to epilepsy and seizures is not as someone who has experienced them firsthand but as someone who lived alongside them, with all the observations, and limitations, that a perspective like this entails.

My sister, Gillian, who was a year behind me in school and with whom I have always been close, was first diagnosed with epilepsy when she was five. I say "first" because her diagnosis likely came years after her earliest seizures and would evolve over time. My own memories of this period are minimal, coming into focus mostly around some disjuncture in our usual, in-step routine: Gillian missing from our shared bedroom at night, doctor's appointments in other cities, and my periodically walking home from school alone. Moving forward in time, the circumstances surrounding her epilepsy become more pronounced in my mind: being pulled from class because Gillian had a seizure, trying but ultimately failing to stay up with her all night in preparation for medical tests the next day, and feeling the relief, or disappointment, of occasionally canceled events. To remember in this way probably reflects not only my age but also how little I actually experienced epilepsy or, in fact, had to think about it most of the time.

By late elementary and middle school, Gillian's seizures became much more frequent and severe. None of the drug combinations she tried seemed to mitigate the tonic-clonic and absence seizures that doctors said came from a region of the brain located just behind her left ear. When her partial tonic-clonic seizures (what were once called grand mal seizures) began to last beyond five minutes and to occur serially, one after the other, I learned how to hold tranquilizer tablets in her cheek until her entire body relaxed and she then slept most of the day. Besides such lengthy seizures, there were those preceded by vomiting, seizures in which she injured herself on nearby objects and furniture, and those that drained all color from her face. She frequently bit the inside of her mouth and tongue while unconscious, broke blood vessels around her eyes, and would later explain that she suffered headaches, muscle strain, and difficulties processing information for days afterward. There were also seizures without convulsions, during which her attention was only momentarily diverted or her language became confused before everything, seemingly, would return to as it had been before.

In the absence of adequate drug options or control, Gillian spent the summer following her first year of high school in an epilepsy-monitoring unit to assess her candidacy for brain surgery. For six weeks she remained in a ten-bed hospital room under twenty-four-hour video and electroencephalographic monitoring, wired, essentially, to an EEG machine, so doctors could learn more about her seizures. After confirming the focal nature of her epilepsy (Gillian's seizures always began in a single region of the brain—one associated with memory, language, and processing of auditory information), her doctors deemed her eligible for surgery.

Gillian's first neurosurgery, that fall, initially seemed successful. The operation removed several grams of cortical tissue from the anterior side of her left temporal lobe. Early in her recovery, however, her seizures returned. While shorter and less intense, these seizures had few warning sensations, making them hard for her to navigate safely. Within two years, she thus repeated the surgery, this time undergoing a shorter but more invasive type of presurgical monitoring. Subdural electroencephalography involved doctors drilling small holes through sections of her skull and placing receptors directly onto the surface of her brain to continuously track its electrical output. This provided more localized feedback than traditional EEG taken from the head's surface, though it posed additional risks. Remarkably, it was only during this period of more aggressive medical intervention in Gillian's life that I remember for the first time fearing, at least consciously, that she might be in some danger or even die.

Gillian's epilepsy, after all, long retained some element of normalcy for me. However intense her seizures became during our childhood, they were regular events that sometimes happened at home, at school, or anywhere else we went together. Though I was habituated to it, her epilepsy seemed, to me, largely undetectable to the outside world; no one knew of Gillian's diagnosis unless she told them or had a seizure. When seizures occurred in public, however, I saw how they produced strong, often negative, emotions in others and transformed nearly everything I had taken for granted about the social environment only moments earlier. Less obvious to me then was just how quickly I moved from these instances of recognition to barely seeing anything of Gillian's experience at all.

Yet, simultaneously, I sensed from an early age that my sister might not be totally safe—both physically and in subtler ways, given the confusion and often palpable discomfort that surrounded her epilepsy. The late twentieth-century period in which we grew up was an era of landmark legislation for disability rights and antidiscrimination in several countries around the world, including Canada, where we lived, and the United States, where I would later go to study.* Even so, it was not uncommon for the adults in our lives to be openly skeptical or dismissive of Gillian's missing school, needing additional time to complete some task, or experiencing considerable side effects from her medications. Similar was a kind of well-meaning hesitancy that sometimes arose around her participation in certain activities both within and outside the family. These features were almost constantly present and became points where I sometimes intervened, though she and I never discussed how any of it affected her at the time.

When I began researching epilepsy as a doctoral student in the history of science, I noticed two things. First and most striking was just how little in the historical literature had been written about epilepsy. Most of what did exist focused on advances in medical understanding and treatment over several centuries or on the intensification of institutional and eugenic approaches by the early twentieth century. There was little beyond that time period or that resonated with my own witnessed experiences rooted in questions of epilepsy's *visibility*, *safety*, and *control*. Second, however, was just how unmistakably the elements of these more contemporary cultural perspectives appeared—and indeed, seemed to originate from within—the source material of the 1930s to the 1950s. That understudied period in epilepsy's history, specifically how epilepsy changed mid-century and what it then meant for people to carry it, is the focus of this book.

I tell pieces of Gillian's story here, then, as much to explain my own connection to the material as I do to highlight a particular lineage. Again and again in my research, I saw a number of familiar themes and framings around epilepsy emerge in sources of the mid–twentieth century. Seizures, for instance, were increasingly depicted as events that revealed an otherwise minimally visible condition to the public—a fairly new construction of epilepsy as imperceptible or "invisible" that has remained in contemporary discussions,

---

*For instance, see section 15 of the Canadian Charter of Rights and Freedoms, 1982, and the Americans with Disabilities Act, 1990.

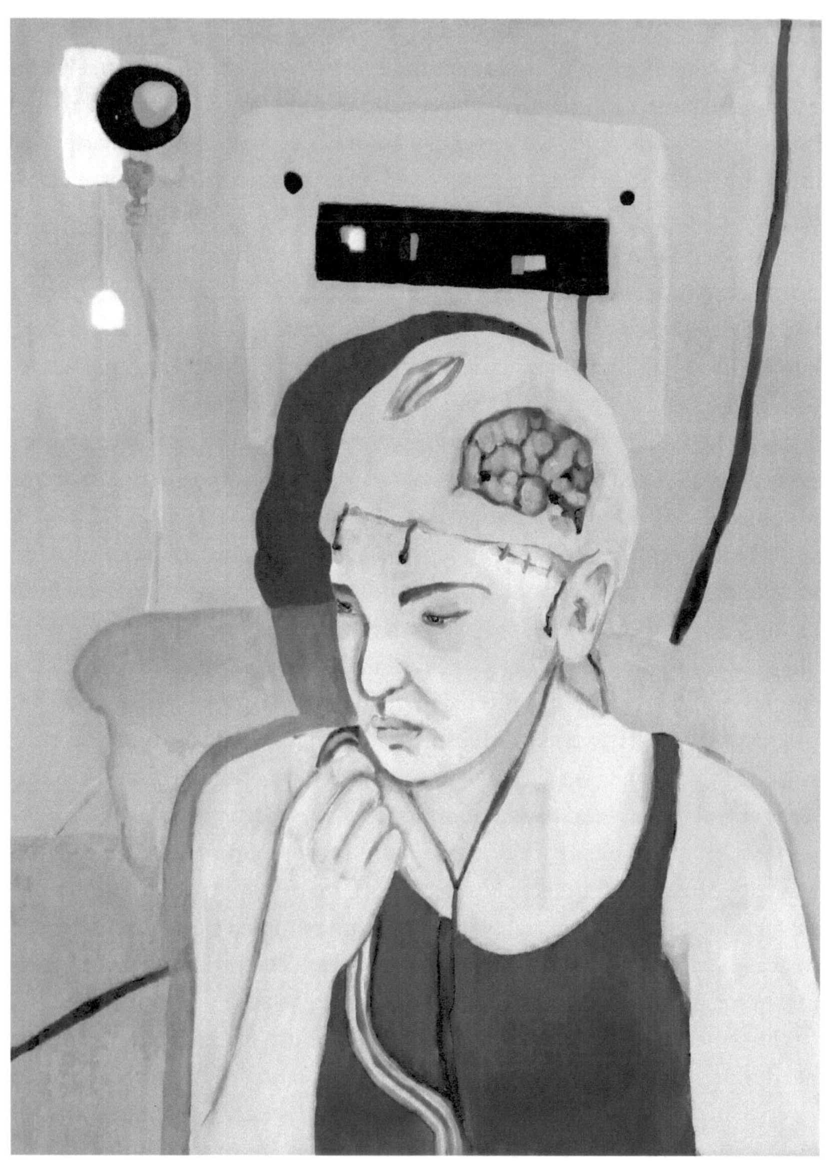

"Epilepsy Unit," self-portrait at London Health Sciences Centre. Painting by Gillian Elder, 2001. Courtesy of the artist

and indeed experiences, of several chronic illnesses and disabilities today. As new and more effective drugs and diagnostic techniques appeared in this era, there were also parallels with more recent decades in the implied capacity of people to control their epilepsy—whether by taking medication or by avoiding seizure-"triggering" behaviors. The further I waded into the project, the more I also recognized early destigmatization efforts that predated the full swing of disability and patients' rights movements—efforts, however, that sought to uplift people with epilepsy by downplaying epilepsy's significance and distancing it from other disabilities. Equally in these sources, I saw the lingering ambiguity of family histories in situating a diagnosis of epilepsy, rising expectations around the management of other people's feelings relative to seizures, and the often complicated ethics surrounding new therapies, such as neurosurgeries and anticonvulsant drugs that we still, in some form, use today. These issues arising from the mid-twentieth-century period examined in this book were mirrored back to me, growing up, a half century later.

Any illness or disability at a given moment and place is a particular kind of internal and social phenomenon. I do not know—then or now—what it feels like to have a seizure, to move through the world with knowledge that I could have one, or to inhabit the label of "epilepsy." I also do not claim that what follows in this book captures the wide array of lived experiences of epilepsies and seizures in the mid-twentieth-century United States—experiences that, as I explain in the introduction to the book, remain untold stories more than I had hoped. At the same time, however, I believe I have some insight into those experiences and existing sources because I lived alongside epilepsy in a single, more recent permutation. Gillian's story is neither mine to tell nor is it directly related to those covered in the book. She has nevertheless in many ways shaped my perceptions as a person and an author. This book is therefore, and in dedication, both for her and because of her.

Montreal, May 2024

Acknowledgments

Many people and institutions helped with the researching and writing of this book. A very special thanks to Whitney Laemmli and Tamar Novick, who read and reread parts of the manuscript with care, as well as to Zain Lakhani and Meggie Crnic, who have been generous supporters and readers of my work over the years. I would also like to thank the community of graduate students and faculty members of the Department of History and Sociology of Science at the University of Pennsylvania—too numerous to name here—for their expertise, mentorship, and comradery of a decade ago and more. Thank you, too, to my original doctoral committee, Beth Linker, Robert Aronowitz, and Kathy Peiss, for their knowledge and support. Though this project started as a dissertation at the University of Pennsylvania, it became a book in the cities and positions that I occupied after leaving Penn; thank you to colleagues at the Program in the History of Science and Medicine at Yale University, the Center for Social Medicine at the University of California, Los Angeles, and my most recent institutional home, the Department of Social Studies of Medicine at McGill University. A particular thanks to McGill's Thomas Schlich for his support of me and my work as a historian.

In addition, several grant and funding bodies made this research possible. I thank the Social Sciences and Humanities Research Council of Canada and the University of Pennsylvania for their doctoral fellowships and research grants. Thank you, too, to the Osler Library at McGill University for fellowships funding portions of this research. I am equally indebted to the many librarians, archivists, and subject experts who provided research assistance in cities across the United States and Canada. Thank you to staff at the Countway Library of Medicine at Harvard University, Walter P. Reuther Library at Wayne State University, Wisconsin Historical Society Library and Archives at the University of Wisconsin, New York Academy of Medicine Library, National Museum of American History Archives Center, Kinsey Institute Library & Special Collections, Montreal Neurological Institute-Hospital, Osler Library of the History of Medicine, Archives of Ontario, Margaret Herrick

Library, and UCLA Film & Television Archive Research and Study Center. Special thanks are owed to the late Dr. William Feindel, former director of the Montreal Neurological Institute and then guardian of the Penfield papers, for allowing me access to a wealth of files and patient records used throughout the book, as well as to the late Dr. Granville Nickerson and Mary Louise Nickerson for funding this research. I would also like to thank my anonymous reviewers and all those who helped prepare the book for its final publication. Thank you to the editors at Johns Hopkins University Press, including Matthew McAdam, Mara Mills, Wayne Tan, Jaipreet Virdi, Robert Brown, and Phoebe Oathout, as well as to Heidi Fritschel and Sergey Lobachev for their assistance with copyediting and indexing.

Finally, thank you to my family, my friends, and to Alexandre and our two dear children.

Abbreviations of Archives

AO	Archives of Ontario, Toronto, ON
CLM	Countway Library of Medicine, Harvard University, Boston, MA
FTA	UCLA Film & Television Archive, Archive Research and Study Center, University of California, Los Angeles, CA
KIL	Kinsey Institute Library & Special Collections, Indiana University, Bloomington, IN
MHL	Margaret Herrick Library, Academy of Motion Picture Arts and Sciences, Beverly Hills, CA
MNI	Montreal Neurological Institute-Hospital, Montreal, QC
NMAH	National Museum of American History Archives Center, Smithsonian Institution, Washington, DC
NYAM	New York Academy of Medicine Library, New York, NY
OLHM	Osler Library of the History of Medicine, McGill University, Montreal, QC
WHSL	Wisconsin Historical Society Library, State Archives, University of Wisconsin, Madison, WI
WPR	Walter P. Reuther Library, Archives of Labor and Urban Affairs, Wayne State University, Detroit, MI

Secrecy and Safety

Introduction

Seizure Prone

Having "lost job after job," a Wyoming man, writing to American neurologist and neurosurgeon Dr. Wilder Penfield, reasoned that his seizures had ruled him out: "When an employer sees me in this condition, I am immediately discharged."[1] Of his time in the United States Army, another letter writer confided, "I thought the soldiers would have a good doctor, but I got kicked out."[2] Desperate for a cure, a woman in Texas wrote, "I just couldn't marry anyone as long as I am afflicted with this," while another in New York described her daughter's rejections from and at school as the source of a "definite complex."[3]

These accounts, all written to Dr. Penfield in 1948, come from a range of people across the United States who were living with epilepsy. Yet few who then had seizures discussed their experiences of exclusion so openly.[4] Deemed a heritable and incurable mental and nervous disease for centuries prior, epilepsy was deeply stigmatizing and prominent within the country's eugenic laws.[5] Marriage prohibitions stood in many states, as did sterilization, upheld by the Supreme Court as recently as 1927.[6] Epilepsy could also lead to dismissals from work and school, restrictions at national borders, and permanent institutionalization. Despite declining beliefs in its contagious or mystical nature, most judged it a painful situation to bear.[7]

During this same mid-century period, however, public attitudes, and indeed the experience of epilepsy itself, were said to be changing. According to magazine and health reports, at least half a million people in the United States had epilepsy, and most of them could be considered "normal," both socially and mentally.[8] Electroencephalography, or EEG, had revealed the brain's electrical activity, making the once enigmatic illness traceable. Re-

Early pamphlet emphasizing public illumination and control. Cover of *Epilepsy—The Ghost Is out of the Closet*, by Herbert Yahraes (New York: Public Affairs Committee, 1944). Courtesy of the New York Academy of Medicine Library

leased to the market in 1938, Dilantin became the first in a host of break-through medications said to control seizures. Delicate new neurosurgeries, like those performed by Dr. Penfield, made complex cases operable for the first time in history.[9] As a result, a growing chorus of experts and advocates, including neurologists and state health organizations, argued that epilepsy should no longer be feared, for most people, they reasoned, could receive modern treatment and experience the life-altering control they had always deserved. Centuries of shame and secrecy, it seemed, would dissipate as quickly as the bafflement over epilepsy itself.[10]

This book examines these dramatic revisions to public depictions of epilepsy. Once considered an all-consuming disease, epilepsy was made anew as an eminently controllable medical condition and even minor disability in the years following World War II. The mid-twentieth-century period is generally characterized as one of avid faith in the strength and authority of Western biomedicine. The celebrated medical advances of the preceding half century, including antibiotics and vaccines, reshaped expectations about the extent to which a number of major illnesses could be understood and even eradicated.[11] This image of control and enlightenment nevertheless belies a more complicated picture. As this project highlights for readers, whether historians, clinicians, disability scholars, bioethicists, or simply those interested in epilepsy, both seizures and social prejudice persisted throughout this era. By the mid-1950s, anticonvulsant drugs like Dilantin had significantly lessened the incidence of seizures: by 80 percent or more in approximately half of all cases, with smaller reductions in another quarter of those treated.[12] Though this was a major gain from previous therapies, the possibility of seizures, much like the diagnosis of epilepsy itself, remained. Anyone with epilepsy in this period of newfound hope and discovery was still essentially seizure *prone*, not seizure *free*. Moreover, as evidenced by obstacles to employment, education, and other rights extending into the 1970s and later, an exclusionary legacy endured. These facts upheld existing barriers and created new ones, each of which animates a still mostly unwritten history.

How a socially fraught and often inapparent illness existed and persisted in this era remains a larger story than the sum of the medical developments newly surrounding it. As this book will illuminate, mid-twentieth-century efforts to render epilepsy controllable in the absence of total control or actual acceptance fostered an unrecognized transformation—one in which ideas about personal and medical control became central to public life and identity—and those who continued to live with the possibility of seizures were compelled to contain themselves in new ways. This work is an effort to find something of that more elusive history.

Seizure Control

Epilepsy is a neurological disorder and disability that today affects an estimated 3 million Americans and 65 million people globally, equaling about 1 percent of the population.[13] Often appearing in childhood and having several known causes and types, it is currently defined by a history of two or more seizures occurring at least one day apart. Not everyone who experiences an isolated seizure or seizures receives a diagnosis of epilepsy, though everyone with some form of epilepsy has a record of recurring seizures that ultimately defines their condition.[14]

Seizures, essentially surges of electrical activity between neurons in the brain, are as diverse as the brain's many functions. Though individuals tend to experience the same kinds repeatedly, they are rarely predictable and can manifest as anything from full losses of consciousness with convulsions, to short gaps in awareness, to involuntary movements of the body with consciousness intact.[15] Beyond their many presentations, neurologists categorize seizures as either focal (also called "partial"), meaning that they begin and remain in one region of the brain, or generalized, suggesting widespread seizure activity in both the brain's hemispheres. While not often dangerous themselves, seizures' rapid onset and impairment of the nervous system can increase the chance of accidents for people with epilepsy. Prolonged, frequent seizures can also lead to acquired brain injuries for some and, in rare but tragic cases, sudden death.[16] Whether someone achieves a high degree of medical control or sees little change with treatment, a diagnosis of epilepsy is usually long term. People with epilepsy thus experience it as a chronic condition; they typically manage medications, lifestyle factors, and the more or less likely chance that seizures will occur.[17]

For many people, however, the truth of epilepsy lies not in medical definitions or control efforts but in what seizures themselves entail. Foregrounding first-person accounts, one landmark study in 1983 characterized seizures as an "assault" and "involuntary body changes that may obliterate one's sense of self control and contact with the world."[18] Interviewees emphasized often substantial sensory changes and the auras preceding their seizures. Others described being "knocked out" or "away" and the sense of helplessness that

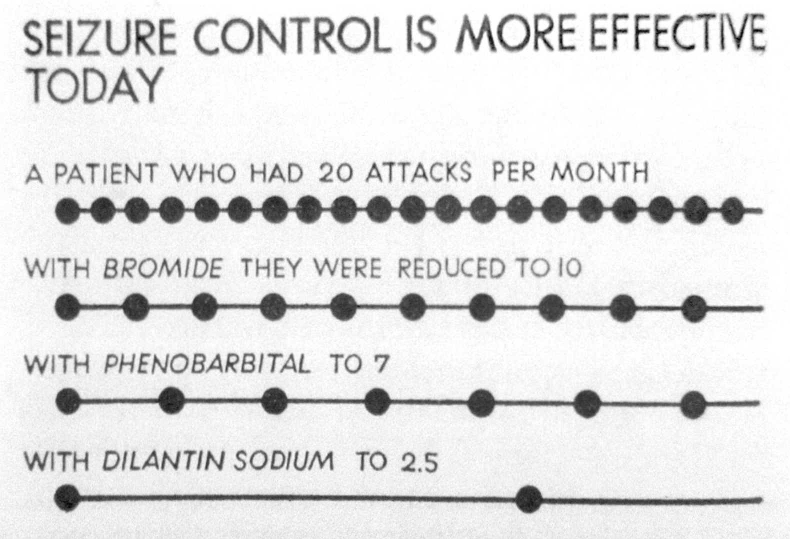

Graphic demonstrating the improved control of seizures with Dilantin. From *Epilepsy—The Ghost Is out of the Closet*, by Herbert Yahraes (New York: Public Affairs Committee, 1944), p. 24. Courtesy of the New York Academy of Medicine Library

followed. As one respondent put it, seizures were "a process of total disintegration from one's whole self," of falling "victim" to "a force outside my power, my control."[19] The "assault" and true meaning of epilepsy therefore consisted of not only a seizure's physical effects—or the context in which it might appear—but its affront to a perceived capacity for control, critical to one's very sense of self and being.

Writing of her second "major attack" three decades earlier, in postwar England, author Margiad Evans described similar feelings of dislocation, inner mistrust, and a diminished sense of self-control:

> Ever since I have been incredulous of all things firm and material. The light has held patches of invisible blackness . . . the sense of being, which is life . . . a thing taken and given irresponsibly without warning. . . . Sight, hearing, touch, consciousness, torn from one like a nest from a bird! . . . Every one [sic] knows what electricity cuts are or a fuse from lightning when every light goes out. Imagine that darkness, and yourself going with the light, to reappear when it comes on again, but doing something totally different.[20]

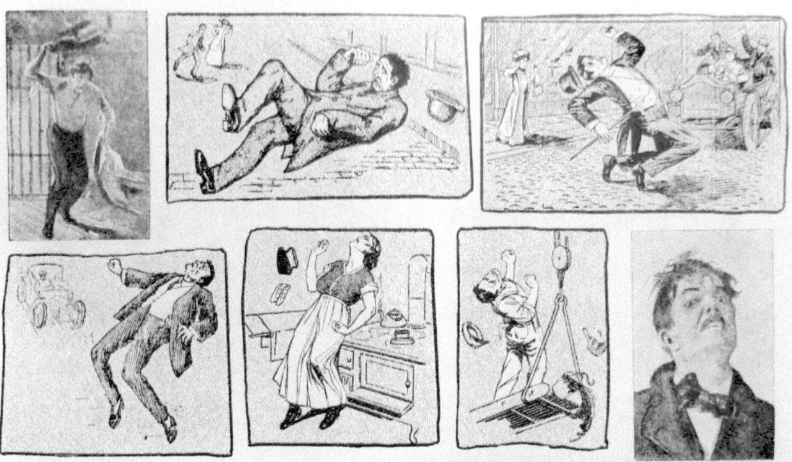

A sample of imagery depicting the visible and public nature of seizures taken from early twentieth-century advertisements for patent medicines. From the pamphlet *Epilepsy "Cures" and "Treatments"* (Chicago: American Medical Association's Bureau of Investigation, ca. 1929), p. 14. Courtesy of the New York Academy of Medicine Library

According to such accounts, the essential experience of epilepsy appears to be the seizure—an event in which one's body upends itself with little or no warning, leading to the physical and social undoing of an otherwise "whole" self. Some seem to have experienced this loss of control as a force "outside" or in spite of oneself, if not antithetical to oneself, and thus as a crisis of personal cohesion and autonomy.

It is tempting to see such perceptions of control and personhood, oriented around seizures, as timeless or even inevitable features of epilepsy. This book proposes, however, that they reflect new perspectives at a particular point in history. What changed around the middle of the twentieth century was not seizures or their nature but the introduction of a powerful assertion that they could be medically controlled.[21] By the late 1940s, US federal and state initiatives to include people with epilepsy in areas of public life, such as education, employment, transportation, and government programs, hinged on these claims. Seizure control promised improved social outcomes. It was also central to public reasoning about the seizure-prone person's place in the postwar social order. That those with epilepsy were "normal" and made further so with medical help became a leading argument of neurologists, health agencies, social workers, educators, parents, and, not least, people with epilepsy. Such arguments also revealed beliefs about the ability, and even duty, of individuals to exhibit control as the clearest expression of their worth.[22]

Epilepsy in the twentieth century thus became a "manageable" and largely "invisible" medical condition and disability, and seizures, the lone, containable symptom and sign. These mid-century understandings inspired new and more public ways of defining seizures and their risks.[23] They also promoted a belief that those with epilepsy should command and conceal their condition to attain the social protections and privileges previously denied to them. The imagined place of control, in effect, became seizure-prone people themselves, while epilepsy, a once chaotic disorder, became disciplined by orderly care.[24] Seizures, in turn, represented not only a loss of physical control or the illness itself but the dissolution of an actual and complete self. In this way, having epilepsy came to suggest both the capacity for control and a lack thereof—one that revealed "independence" and "freedom" from seizures as criteria for certain kinds of belonging in American society, while less readily acknowledging the difficulties present in these very terms, or indeed in "control" itself.

Graphic providing statistics of employability of those with epilepsy. From *Epilepsy—The Ghost Is out of the Closet*, by Herbert Yahraes (New York: Public Affairs Committee, 1944), p. 39. Courtesy of the New York Academy of Medicine Library

Between Seizures

A central method of this work is to consider what it means to be *seizure prone*, to imagine epilepsy as an experience characterized not only by either seizures or control but equally by periods of absence and expectancy. By the postwar era, a growing number of people living with and observing epilepsy described it in this way—as a sporadic and therefore *invisible* illness whose defining reality was less about the occurrence of seizures than their ever-present *possibility*.[25] Those writing to Dr. Penfield frequently remarked that they never knew "when or where" a seizure would occur.[26] Their accounts captured dismay at seizures as unwanted and potentially revealing incidents. They also revealed the extent to which living with epilepsy encompassed a broader state of apprehension and uncertainty, if not a growing separateness from seizures altogether.

Before invisible disability and illness became widespread concepts, terms like "part-time handicap" and "invisible" were used by Americans mid-century to describe the personal and relational aspects of seizure latency and risk.[27] Sometimes presented as advantageous features of the experience, these terms referred to the periods between seizures in which one might not appear to have epilepsy. As often, they described detrimental pressures associated with having a marginalized disability only "part" of the time. Paradoxically, both characterizations reflected a tendency to regard seizures as the real or whole experience. Unlike being "epileptic" in the prior sense—an identity label that advocates attempted to shed in this period—being seizure prone marked the absence of but capacity for serious illness. Such ideas were also bound up with notions of control as treatments improved and the impetus to separate seizures from oneself grew. To be seizure prone thus encompassed having an "invisible," "part-time," and even "controlled" illness and disability as much as it reflected a changed understanding of epilepsy itself.

The perspectives outlined here, both related and unrelated to new possibilities for medical control in this era, have not been explored by historians. However, those investigating stigma and deviance concepts beginning in the 1970s exposed elements of this reality early on.[28] As sociologists Joseph Schneider and Peter Conrad describe, "Although epilepsy is stigmatized, except for seizures it has low visibility as an illness." They continue: "People with epilepsy appear to be (and by some definitions are) 'in good health' and lead conventional lives. . . . They must nevertheless continue to manage aspects of their illness."[29] Not only are seizures "'surprise events' that can occur at any moment, disrupting life and causing embarrassment," but they further connote "diminished social competence" that leads people to "control chronic illness in their everyday lives."[30] Postwar accounts show similar readings of epilepsy's chronicity and low visibility and of the need for control, as do those of today.[31] The purpose of the study at hand is to understand how these interrelated elements took shape and, moreover, why people began to see seizures as isolated impairment events at odds with themselves. How did control—with the goal, or at least the consequence, of promoting invisible illness—become an integral and embodied part of epilepsy around the mid-point of the twentieth century? What did it require of seizure-prone people, and what did it create in the culture around them?

Throughout the twentieth century, epilepsy remained present in a percentage of the US population, though far from public awareness. Contrary to some interpretations of the post–World War II period, this was not because most people with epilepsy were institutionalized or controlled the disorder through medication or surgery. Estimates prior to the uptake of anticonvulsant drugs, for instance, suggest that approximately 90 percent of Americans with epilepsy lived outside institutions, and most, regardless of therapy or setting before or after this time, continued to live with some degree of seizure potential.[32] Concurrent with postwar depictions of epilepsy as a temporary, less essentializing impairment was an expectation that those with epilepsy would keep the diagnosis private.[33] Likewise, though doctors and patient groups presented seizures as safe and manageable occurrences in this era, seizures were simultaneously designated as unchecked threats on roads and in workplaces, leading to a host of new laws and oversight.[34] In effect, as people with epilepsy were permitted and even encouraged to participate in new public spaces after World War II, they also became responsible for maintaining strict standards of control and for delivering impossible assurances that seizures would not occur.

Secrecy and safety are thus ubiquitous themes in this book. They describe a range of adaptations that grew from a postwar paradigm of seizure control, reflected in everything from formal public policies and discourses to modified social codes and individual behaviors. Practices of secrecy and safety cohered amenably around epilepsy, since many postwar actors deemed it a stigmatized but increasingly invisible illness and identity, easily delimited to seizures alone. Prior to the rise of a broader disability rights movement or later-in-the-century efforts to destigmatize various illnesses, such as mental illnesses, the concepts of secrecy and safety also explain how control became an issue of personal maintenance and the fact of seizure potential remained mostly unseen. Though seizures would in theory become discrete, controllable phenomena, it was ultimately their absences, as much as seizures themselves, that required management in a new public light.

Histories of Uncertain Belonging

This book contributes to connected, though not always well aligned, histories of medicine, health and the body, and disability. Most plainly, it expands a sparse literature on the history of epilepsy itself. To date, no book-length studies of epilepsy in the twentieth-century United States exist, nor are there many shorter works framed chronologically beyond the early interwar period.[35] Most works also tend to focus on key medical developments, spiritual elements in premodern histories, and medical and eugenic institutions in modern ones.[36] Owsei Temkin's foundational work in 1945 first showed how scientific explanations of epilepsy in the mid-nineteenth-century West replaced long-standing religious and supernatural theories, marking the "end of the falling sickness" and what sociologists later identified as the beginning of its medicalization.[37] Historians of medicine who examined the period of the 1860s to World War I later focused on how a rising profession of neurologists further delineated epilepsy's true form.[38] Others, such as Ellen Dwyer and Dea Boster, have since explored the effects of the eugenic and social frameworks of the nineteenth and early twentieth centuries, as well as epilepsy's place in violence and resistance within American slavery.[39]

Besides extending the chronological scope of this literature, this book argues against a wider vision of medicine as illness's logical resolution, if not the central force driving its story. It simultaneously challenges implicit suggestions that the primary experience of most people with epilepsy before and during the mid–twentieth century was a sequestered or institutionalized one rather than one that was, at least in part, public in nature.[40] Both primary and retrospective works tend to attribute "modern" and "enlightened" attitudes toward epilepsy to near concurrent advances in anticonvulsant drugs, electroencephalography, and neurosurgery beginning in the 1930s.[41] Even today, the idea persists that a reduction in discrimination against people with epilepsy occurred because epilepsy became "controlled medically," leaving the more uneven effects of these changes unexamined.[42] Missing are studies that address the complexities introduced by new drugs and diagnostics, akin to the work some historians of medicine have done on the impact of insulin in managing diabetes or on the ambiguities of visualization technologies in representing disease (see chapters 1 and 2).[43] Such analyses show not only how medicine transforms illness or holds it to the light but also how patients and stigma shift around these developments; rarely has prejudice vanished as a result of medical change alone. Equally absent are efforts to understand epilepsy beyond standard disease categories and medical institutions.[44] Disability scholars and illness memoirists in particular emphasize the importance of accounts rooted in the daily and the personal. This book explores these elements through its attention to public space and the meanings of a controlled body at a specific point in time.

Epilepsy's designation as both illness and disability throughout this book may nevertheless create dissonance for some readers. Discussions of the "medical model" within disability studies refer to a view of disability as an individual, real form of impairment requiring biomedical intervention, much like conventional understandings of disease. The "social model," on the other hand, understands disability as a reflection of systemic barriers that overlook and deny people's diverse needs.[45] As some scholars note, however, not all examples fit neatly into these definitions, nor do they necessarily remain static.[46] Epilepsy's earliest classification as a "disability" grew from World War II rehabilitation efforts, yet the simultaneous use of terms like "seizure disorder" in this period continued to imply medical frameworks of understanding.[47] Furthermore, whether "disability" is a term claimed by historical actors or regarded in a purely analytical sense by those examining this history, the role of medicine itself remains a complicating factor. Several people in the postwar era pursued medical care and freedom from seizures, often as they sought to dismantle social prejudice, and for many, the impairment caused by seizures loomed larger. Medicine has therefore been difficult to dismiss as an important aspect of many people's experiences. Though new treatment paradigms and continuing stigma since the postwar era may in part explain these facts, they arguably reveal elements of epilepsy less easily assigned to an illness or disability category alone.

This book's focus on invisibility further enriches disability studies scholarship, which has traditionally explored issues of visibility. Disability scholars first theorized the characteristic visibility of disability in the 1990s, articulating how medical and public acts of looking themselves construct the category of otherness that society calls "disability."[48] More recent work on the invisibility of disability, rather than its intrinsic visibility, detail the effects of such looking turned inward, including the public-private tensions that individuals can face in presenting as "healthy" or "able-bodied" and their frequently alienating experiences of unseen difference.[49] A closely related literature on "disability passing" shows how people with a variety of less overt physical and neurocognitive differences direct their appearance and behavior to avoid stigma and gain benefit.[50] In cases of illness, the concept of invisibility usually refers to chronic illnesses or symptoms falling outside known disease labels, signaling that one's symptoms are doubted or unrecognized by the medical establishment and society. Examples include chronic fatigue syndrome and, more recently, the various symptoms seen in long- or post-COVID conditions.[51] Seizures and seizure proneness share many of these different elements of invisibility. They are also experiences related to but separate from epilepsy in its more commonly understood, and indeed strictly medical, connotations.

Despite a growing literature on invisibility, however, most studies to date have applied the term retrospectively or, more often, used it to characterize exclusively contemporary examples. A historical study of an invisible illness or disability so described by those living with or around it in the past is mostly undocumented.[52] In addition to showing one modern illness and disability's evolution, epilepsy is a notable case given its rapid metamorphosis. Though it became "invisible" by the mid–twentieth century, epilepsy had previously been believed to be quite visible. In nineteenth- and early twentieth-century science and literature predating the focus of this book, people with epilepsy had distinct physiognomies: the effects of chronic seizures, whether pale complexion or bitten lips, marked those afflicted, while behaviors like rage and impulsivity exposed "epileptic personalities."[53] Yet by the post–World War II period, the belief that medical control both disguised and diminished epilepsy converged with already changing notions about its place in the public sphere. Eugenics and World War I–era data showed that relatively few people with epilepsy were institutionalized and heightened public fears that epilepsy was a "hidden" illness. New narratives mid-century suggested that epilepsy might not be discernible beyond seizures and, with some seizures, might not be visible at all, thereby boosting impressions that those afflicted could cross national and state borders, enter the workforce, and go about their lives undetected.[54] Given these changing perceptions of epilepsy's visibility over time, this study invites questions about how other invisible illnesses and disabilities have been constituted and how individual experiences in turn are shaped by them.

Control and self-management concepts are also evident features of social citizenship in this case—what theorists of the body like Susan Wendell have in other instances described as modern myths of medical mastery and bodily control.[55] They are likewise linked to questions of public visibility and inclusion.[56] The postwar period was characterized by increased social and political anxiety in the United States, Cold War concerns about national security, and the rise of the civil rights movement. It was also an era of growing affluence for some, new forms of mass culture, and stringent social norms.[57] Before the disability rights, second-wave feminism, and gay liberation movements came into their own in the 1960s and later, ethics of personal and medical control seem to have reduced the perceived dangers of some seizure-prone people, who could be "regular citizens" provided that their seizures were "controlled." [58] Respectability and comportment strategies such as these were prevalent mid-century where matters of identity and social rights were concerned. The homophile and civil rights movements of the mid-1950s, for instance, demonstrated the utility of one's dress and bearing in contesting inequality.[59] Like parallel campaigns for disabled veterans and polio survivors of this period, such approaches also reflected wider sentiments that people should rise above or "overcome" illness and disability, both on principle and as a public good.[60] The reinforcement of traditional gender roles, suppression of homosexuality, and increased racial oppression are well-known instances of homogeneity that rapidly took root across multiple facets of American culture in this era. Similar trends describe how epilepsy and other stigmatized illnesses, too, were attenuated and contained.[61]

Indeed, for most individuals, the more positive public representations of epilepsy forged in the post–World War II period failed to translate to positive public identities or a more flexible view of control. The English-Welsh writer Margiad Evans, for instance, saw the seizures she developed later in life as evidence that her "security was gone," a statement that evoked a stark social outlook as much as it did any physical reality of her body.[62] Although the earliest US initiatives for people with epilepsy to work, drive, marry, and attend school appeared at this time, these efforts did not generally embrace seizures or promote outward identification by those who had epilepsy (chapters 1 and 5).[63] Moreover, while appeals to "normalcy," "safety," and "control" afforded some Americans limited new opportunities during this era, the entirety of this vision was not available to everyone. As perceptions of epilepsy changed from disease and deviance to treatable medical condition and disability, so too did public representations of seizure-prone people and any opportunities typically advanced the interests of white, male, and mostly younger individuals (chapters 3 and 4).[64] Controlling seizures, in short, unlocked aspects of postwar society for some Americans and left barriers firmly in place for others, including Black Americans, girls, women, older and rural Americans, and those with multiple disabilities.

The remaining fact of seizures also continued to limit seizure-prone people in more encompassing ways. On some level, seizures and their risks troubled not only mid-century American definitions of normalcy but also normative experiences of autonomy and freedom central to political and social ideals of the time.[65] Though the fact of seizures perhaps more accurately highlighted that control as such was illusory for anyone, this perspective was rarely acknowledged. Similarly unrealized was a latent notion that true autonomy and freedom might lie in relinquishing the very idea of personal and physical control. Efforts to certify the safety of some seizure-prone drivers, for instance, failed to reckon with the larger issue of car crashes caused by factors other than seizures (chapter 4). Postwar experts examining the seizure-prone brain as a way of understanding female sexuality likewise pathologized more than they liberated (chapter 2). Rather than embracing these contradictions or the limits of control, the public and people with epilepsy were urged to regard control as a reachable standard and even a norm.

Organization and Sources

The chapters ahead center on the period from 1930 to 1960, examining how a variety of actors understood and managed not just seizures but their ubiquitous possibility in contexts of imperfect control. Chapter 1 traces epilepsy's transformation as an invisible illness and disability over the first half of the twentieth century. It argues that public efforts to improve epilepsy's image mid-century also promoted anonymity; compelled to present as seizure free, people with epilepsy were ultimately constrained by new discourses of medical mastery and control as much as they were elevated or made visible by them. Chapter 2 explores the impact of electroencephalography, a medical technology that read previously invisible bioelectrical activity in the brain and quickly became an authoritative tool of diagnosis. As control efforts grew and dysrhythmic brain waves continued to be found in people who did not have seizures, however, EEG ultimately lost some of its initial power and appeal. Chapter 3 focuses on a Depression-era school for seizure-prone children in Detroit, where after World War II educators removed a variety of seizure-assistive technologies and supports. This was not in response to a decrease in the number of seizures at the school but rather because of a belief in some children's relative "safety" and "normalcy" and thus their suitability for regular schools.

Chapter 4 examines evolving driver's licensing laws. First deemed public dangers in California in 1939, drivers with epilepsy were accepted less than a decade later as safe road risks in several states across the nation. Starting in California and tracing the subsequent rise of medical review boards in Wisconsin and elsewhere, the chapter considers the criteria by which doctors and legislators conferred driving rights upon some seizure-prone people. Those approved gained an ultimate symbol of freedom, as well as a range of control-based risks and responsibilities. Chapter 5 explores the records of American patients, like those quoted at the beginning of this introduction, who sought treatment with Wilder Penfield at the Montreal Neurological Institute. It considers lived experiences of epilepsy at a time of expanding drug therapy and some people's choice of invasive surgery in efforts to control their seizures. It also shows how such patients' unprecedented participation in electrical mappings of the cerebral cortex made critical, though largely invisible, contributions to landmark studies of the brain. The epilogue then explores how the values and strategies associated with a mid-century project of seizure control persist in some form today, despite greater disability and stigma awareness. It simultaneously challenges, however, the notion that control is a fundamentally problematic want reflecting an inherently medicalized view of neurological differences. For many, then as now, recognizing society's shortcomings in supporting seizure-prone people has coexisted with the desire to control and limit seizures.

As a final note on sources, tracing what is hidden can derail the best-laid intentions for writing representational histories. As mentioned, few firsthand accounts exist for this period, and what can be found is often authored by institutions.[66] Given the priorities and prejudices of postwar epilepsy advocacy efforts, available perspectives also lack diversity. Epilepsy, moreover, is often discussed in sources and throughout this book with little differentiation between the array of sub-diagnoses, seizures, and symptoms that individuals experienced. More work remains to be done to trace the encounters of a range of seizure-prone people and to understand how factors such as race, class, gender, sexuality, age, ethnicity, geography, and types of epilepsy have shaped this history.[67] Epilepsy and seizures are likewise named throughout the book in several ways: as illness, disease, disorder, handicap, dysrhythmia, and disability. While some sought to banish the term, "epileptic" was also used mid-century, as were people-first configurations of "person *with* epilepsy" or "person who *has* epilepsy." I have used these terms with as much specificity as possible, as much to reflect their embedded meanings in the sources as to capture an evolving idea and experience.

I VISIBILITY

1

"Invisible Handicap"

From Family Skeletons to Closets

of Near Normalcy

In the 1940 medical melodrama *Dr. Kildare's Crisis*, Mary Lamont, a nurse at Blair Hospital, learns that her older brother, Douglas, has epilepsy. Struggling to come to terms with the unspeakable word and bewildering news, the young nurse at first rejects the diagnosis. Incredulous that her sibling is afflicted, she reasons, "A man who knows he has a thing like that obeys doctor's orders . . . and here was Douglas, out in the city."[1] Mary's disbelief, however, soon turns to dread. Above concern for Douglas, whose decline seems certain, is the question of what the exposed "stain on her family blood" will mean for her upcoming marriage to the handsome physician Jimmy Kildare. Lamont fears various scenarios: the scandal of a broken engagement, the horror of spreading the "taint" to her offspring, and, equally, the unhappiness of a childless marriage. Increasingly dismayed, the nurse plots her escape. Leaving her job and fiancé without warning, Lamont buys two one-way tickets on a ship bound for South America. As she faces a new life dedicated to spinsterhood and the care of an infirm brother, only the knowledge that her sweetheart has been spared and that she and Douglas will soon be hidden comforts the nurse, who believes she is narrowly averting disaster.

Years after the release of the popular film and novel *Dr. Kildare's Crisis*, the *Saturday Evening Post* presented another portrait of epilepsy at the crossroads of public and private life. Leading with a tale of rebuffed patriotism in 1953, one article recounted the formative wartime experience of a New Jersey man who, not unlike the fictional Lamont, had once sat in transit, weighing a secret. On an ordinary day at the height of US mobilization in World War II, the article said, Herbert Jackson endured the censure of a fellow passenger on a city bus. When asked why "a perfectly healthy-looking specimen" such as himself was not in uniform, the young man offered the truth: he had attempted to enlist but was refused because he had epilepsy.[2] As the passenger recoiled, leaving Jackson with the familiar sting of rejection, he resolved never again to volunteer word of his condition. Instead, when such instances arose, he cited ulcers or heart trouble. Such "white lies," the *Post* suggested, were "much less embarrassing" than fact.[3] They were also part and parcel of what Jackson, in his thirties by the time of the article, described as his "invisible handicap."[4]

Despite their differing representations and outcomes, these episodes in early World War II and postwar culture share a central tension: namely, that epilepsy may not initially be seen but stands to become known. Instead of seizures, it is epilepsy's quiescence—the times between seizures when the illness goes undetected—that proves significant. Rather than the drama of the public seizure or the presence of the known "epileptic," it is questions about who knows and might find out that characterize a condition both silent and silenced in American society. At mid-century, epilepsy was largely invisible—it was a condition defined in relation to the public by loaded absences and anxious omissions. This was epilepsy as imagined from both inside and out.

Yet what such narratives meant and how they operated also changed between the years that produced the fictional fleeing siblings in *Dr. Kildare's Crisis* and the real-life rejected recruit featured in the *Saturday Evening Post*. In the short time between the two publications, epilepsy—as an illness both difficult to see and sometimes concealed—transformed rapidly. Throughout the interwar period, epilepsy's low visibility was presumed to be part of its malignance, a function of its nature according to eugenic models of genetic inheritance and institutional care. As an incurable and degenerative disease that ran amok in family bloodlines, epilepsy posed threats beyond those identified as epileptic and remained unsuitable for public discussion.[5] By the postwar era, however, epilepsy's newfound "invisibility" assumed a more public, individual-centered, and even hopeful form. According to experts who sought to destigmatize epilepsy, seizure-reducing drugs and science challenging epilepsy's heritability had made seizure-prone people "treatable" and "normal but at time of seizure."[6] Epilepsy's invisibility beyond seizures thus became an avenue to acceptance and integration, evidence of one's fitting into society rather than an allegory for intrinsic danger.[7]

This chapter explores these changing constructions of invisible illness and disability, tracing epilepsy's meanings as an only periodically and potentially visible condition by the early postwar period. It also considers the contradictions of these developments for seizure-prone people.[8] As epilepsy's visibility in medical and popular discourse increased throughout the postwar era, its presentation in individuals was ideally minimized. Integral to mid-century destigmatization efforts were new principles guiding how seizure-prone people were meant to live and identify relative to their epilepsy, a condition deemed less defining of "near normal" persons than ever before. Popular depictions of epilepsy were gradually freed from notions of family disease or the closely guarded secret and increasingly presented a minor disability that scarcely bore mentioning. This created new meanings around both epilepsy and seizures—particularly for those who continued to live with the diagnosis.

The chapter begins by examining the beliefs and structures that preceded this twentieth-century transition, looking to *Dr. Kildare's Crisis* and its nonfictional parallels in the culture of the Progressive Era and interwar America. It then shifts to explore the changing social and scientific contexts of the late 1930s and early 1940s, including how early doctor-patient groups, like the Laymen's League Against Epilepsy, began to speak publicly about epilepsy following the introduction of new medical technologies and treatments. Finally, the chapter examines an array of postwar newspaper and magazine articles in which the first depictions of epilepsy as invisible were featured. Placed alongside some of the few reported experiences of people with epilepsy in this era, these sources illuminate changing narratives about epilepsy, less easily charted through the social history. They also demonstrate the extent to which the normalizing impulses of postwar culture and a model of open secrecy reshaped visions and experiences of epilepsy in ways usually attributed to medical developments alone.[9]

Proto-Invisibility: Portrait of a Family Secret

In 1940, it was possible to suspend a diagnosis of epilepsy for seventy-five darkened minutes and more than one hundred serialized pages with considerable dramatic tension.[10] Appearing in film and in *Argosy* magazine during December 1940, *Dr. Kildare's Crisis* capitalized on the connotations of a mostly unspoken word. The story was the fifth in an already popular book series named for its title physician and written under the pen name Max

Brand by Fredrick Schiller Faust. Among the first in the series adapted into a motion picture with MGM Studios, the movie featured an illustrious cast, including Lionel Barrymore as the cantankerous senior internist Dr. Gillespie.[11] By the following year, *Dr. Kildare's Crisis* was published as a full-length novel by a trade publisher that produced a total of seven Kildare books, including *Young Dr. Kildare* (1938) and *Dr. Kildare's Search* (1943). Shortly thereafter, the prolific Faust, best remembered for his pulp westerns, died in Italy while serving as a war correspondent for *Harper's* magazine. At the time of his death, Faust had nearly twenty recognizable pseudonyms but was virtually unknown to the public. The popularity of Max Brand, however, was such that American soldiers' kitbags continued to be outfitted with pocket-sized editions of Kildare favorites, including *The Secret of Dr. Kildare*.[12]

Establishing what would quickly become formula in medical and hospital genre fiction, *Dr. Kildare's Crisis* enlisted a mode of detective diagnosis to cast its narrative arc. Like a private investigator tracking a suspect, Jimmy Kildare—a consummate "hunting dog" and "miracle man" who can "read our secrets at a glance"—inches toward the truth of his soon-to-be brother-in-law's strange constellation of symptoms.[13] Upon meeting Lamont, Kildare intuits that there is something off about his future relation. The doctor first notes his "starved body, inadequate to the bearing of burdens" and "eyes from which the light had been burned out."[14] With each subsequent meeting, he observes more of the man: his fragile, gaunt appearance and the "hypnotized, sleepwalker look in his eyes."[15] Particularly unsettling to Kildare are Douglas's eyes, which at times seem devoid of life, "utterly emptied, like the colored glass from which sparkling water has been poured."[16] As disturbing to the doctor as his physical attributes, however, is the way Douglas periodically "burns up" with energy. He goes from obsessive and grandiose—proclaiming "I cannot fail . . . I feel as if God had commissioned me"—to scarcely able to converse the next day.[17] Douglas's talk is also distinctly socialist, with schemes involving communal land and the rejection of modern technology.[18] Yet just as quickly as he becomes animated, the "man on fire" becomes dull and confused. With these oscillations, Kildare is convinced that there is "something pathologically wrong with this man . . . so wrong that he regretted kinship with Mary."[19]

Feeling Kildare's cooling affections almost as much as his scrutiny, "as if a detective had been probing at her family secrets," Mary too seeks to understand the curious developments in her brother's behavior. Ignoring warnings

Movie still from *Dr. Kildare's Crisis*, 1940. From left to right: Robert Young (Douglas Lamont), Laraine Day (Mary Lamont), Lew Ayres (Dr. Jimmy Kildare), and Lionel Barrymore (Dr. Leonard Gillespie). Core Collection Production files of the Margaret Herrick Library, Academy of Motion Picture Arts and Sciences

from a cousin that "family skeletons" are "best kept in a dark closet," she learns through covert correspondence with another relative of an uncle's mental illness and death in an institution. For Mary, the discovery is the "drop of poison enough to stain the record of a whole family," as well as to explain Douglas's state.[20] Meanwhile, Kildare, who knows nothing of this family history, though something of the growing secrets between himself and his bride, plans a clandestine medical experiment. Over the course of a heavy dinner of steak and beer in the cover of Douglas's room, he confirms his hitherto unnamed suspicions of epilepsy. When Douglas rises from the table after the meal, he collapses to the floor in a series of convulsions that progress violently into the night.[21] With Douglas's shadow side fully exposed, the couple's future together falters, driving Mary to flee on the *Caballero*.

Though a sensational work of fiction, *Dr. Kildare's Crisis* reveals elements essential to understandings and experiences of epilepsy in the United States in the years approaching World War II. Throughout the early decades of the twentieth century, epilepsy had been considered at least doubly hidden. Its presumed hereditability, for one, meant that it was concealed in a biological sense.[22] As with the initially asymptomatic Douglas Lamont, seizures might be slow to present themselves but were believed to reside in one's genetic makeup, connected closely perhaps to the mental illness or moral vices of near and distant relatives. This rendered seemingly healthy family members, like Mary Lamont, witting or unwitting bearers of the disease.[23] Second, as a thoroughly stigmatized condition linked to various ills, epilepsy might be intentionally obscured. Hence, Douglas, if he realizes he has seizures, never acknowledges them, and Mary, who is implicated in her brother's illness, banishes herself when she sees no way to withhold the information from her fiancé or spare their future offspring from epilepsy.

Epilepsy more broadly throughout this period was regarded as a degenerative mental and nervous disease with few promising treatments.[24] Characterized by physicians as a pernicious "habit" of the nervous system, seizures were considered both unseemly and dangerous, producing an inevitable deterioration of body and mind. Kildare thus reasons that "nerves are like shock troops" and that "Douglas was using them day and night."[25] His rapid decline only strengthens the doctor's impression of epilepsy as a "serious and progressive weakening of the brain" in which "the mind is apt to shrivel and die away."[26] Simultaneously, epilepsy had long-standing associations with mystical genius. Presented as something of a visionary, Douglas has insights that appear following his seizures, evoking ancient beliefs in the divine and spiritually possessed epileptic alike. The peripheral mad uncle, Douglas's erratic behavior, and Mary's heightened nervousness in first anticipating then observing seizures collectively advance themes of mental and familial unraveling. Yet it is the book's closing portrait of a man ravaged by seizures and removed from public view that offers the final, bleak outlook for the seizure sufferer and his kin. As shown by the divergence between the novel and the film (the latter only narrowly cleared censor boards in 1940 despite its revised ending of a curative brain surgery for Douglas that allows the sweethearts to marry), epilepsy remained a topic suffused with fascination and fear.[27]

That epilepsy ran in families was a well-accepted fact during the period that produced Faust and his memorable Kildare series. Understandings of a transmissible family illness extended back centuries but gained further strength with eugenics. Emerging from nineteenth-century British genetics and widely influential in the United States by the early twentieth century, eugenics, as a social and scientific movement of "race betterment," focused on promoting "better breeding" in humans.[28] This included minimizing the spread of heritable "defects" and disease. In efforts to preserve the biological integrity of future generations, eugenicists designated people with epilepsy—along with poor white Americans, African Americans, persons of lower measured intelligence, and those with mental illnesses and physical disabilities—as among the most unfit to reproduce.[29] Not only was epilepsy an undesirable trait, but eugenic family studies of the period demonstrated the extent to which it commingled with insanity, criminality, alcoholism, physical "deformities," and "feeblemindedness."[30] Although epilepsy was thought to have many causes, idiopathic and constitutional epilepsy—as opposed to traumatic or acquired kinds—came to represent the largest classification group by the early twentieth century, reinforcing popular and scientific notions of its genetic basis.[31]

World War I military statistics similarly supported perceptions of epilepsy as a spreading concern in nearly all segments of the population in this era. In what soon became a widely cited figure, reports from the surgeon general in 1917 found that between 2.7 and 5.23 of every 1,000 men screened by the US Army had epilepsy.[32] Such numbers caused consternation about the unseen scale and scope of the condition, stoking anxiety among white experts about race and epilepsy in particular. In 1923, Charles Davenport, director of the Eugenics Record Office at Cold Springs Harbor, noted that while the ratio of epilepsy among Black recruits in World War I had indeed been higher than in white recruits, African Americans were slightly less affected by epilepsy overall.[33] Contradicting nineteenth-century wisdom that suggested a higher incidence of epilepsy in Black populations, particularly in the South, twentieth-century eugenicists found that the predominantly white and wealthier New England states had the highest, albeit modestly higher, rates of epilepsy. Though much eugenic fieldwork of the period focused on poor, white, institutionalized Americans, the prevalence of epilepsy in the homes of white professional and learned families was noted with alarm.[34]

Eugenics laws and institutionalization increased epilepsy's stigma and concealment throughout this period. Beginning with Connecticut in 1895, seventeen states placed limitations on marriage licenses, restricting those with epilepsy from marrying and making epilepsy excusable grounds for divorce. In 1903, the immigration of people with epilepsy to the United States was prohibited by law.[35] Following marriage and immigration laws, several states passed sterilization statutes, including Indiana in 1907, California, Connecticut, and Washington in 1909, and others thereafter.[36] In 1927, the US Supreme Court's ruling in *Buck v. Bell* upheld the practice with its infamous decision regarding Carrie Buck, a teenaged inmate at the Virginia State Colony for Epileptics and Feebleminded, with the proclamation that "three generations of imbeciles are enough."[37]

People with epilepsy were also committed to private and state asylums for the insane throughout the nineteenth and early twentieth centuries (epilepsy was closely linked to hysteria in the nineteenth century) and housed in what were called epileptic colonies or farms in growing numbers beginning in the 1890s.[38] Among the earliest of these colony institutions was the Ohio Hospital for Epileptics, founded in 1893, and the Craig Colony, established in upstate New York in 1906. Like their psychiatric counterparts, institutions of this kind were both a symbol and a means of segregating people with epilepsy from the community. They also frequently grouped "epileptic" and "feebleminded" individuals together given a presumed lack of treatability and the overlap believed to exist between the populations.[39] Whether deemed inherent or the effect of recurring seizures on the brain, the "low mentality" of the "epileptic," like that of the "mental deficient," was noted. So, too, was the apparent ease with which such issues seemed to pass freely through family bloodlines.

Epilepsy and Family Life

Despite the expansion of eugenic laws and turn-of-the-century institutions for epilepsy, evidence suggests that most people with epilepsy resided at home and within their communities during this era. Only one in ten people with epilepsy are estimated to have been institutionalized during the first three decades of the twentieth century, making it likely that most cases remained unknown or concealed to some degree.[40] Institutionalization, a less prominent but significant element of epilepsy's history, mostly becomes apparent retrospectively, and in fragments, among more eminent Americans. Author Mark Twain, for example, sent his youngest daughter, Jean Clemens, to a sanitarium in Katonah, New York, after her grand mal seizures worsened following a childhood head injury. As Karen Lystra details, Jean's separation from the family was both part of her advised treatment and a means of removing stress from the family.[41] Her placement there, as with others who were in her social class and who perhaps initially had fewer seizures, was considered temporary and justified on the grounds of its therapeutic value. Twain wrote publicly of Jean's condition for the first time after she was found dead at age twenty-nine from the effects of a seizure soon after returning home for Christmas in 1909. Her longer struggle with epilepsy and separation from the family for nearly three years, however, remained private for decades thereafter.[42]

As a family affair, efforts to disguise epilepsy in the early twentieth century probably took place from a person's childhood. This was when, aside from traumatic cases, seizures most commonly appeared and when most individuals remained under the care of parents or extended family. Though both parents were implicated in the health and fitness of a child, mothers were usually considered more likely to transmit illness and disability to their offspring.[43] Enduring beliefs that shock and fright during pregnancy caused seizures increased maternal stigma, as did the illnesses and injuries of birth and infancy known to produce epilepsy later in a child's life.[44] In two-parent households where the father and not the mother worked outside the home, it is also probable that mothers assumed greater duties of maintaining the appearance of family health alongside those of child-rearing and caring for family members. Children who had seizures at school, for instance, were typically sent home and unable to return.[45] At the same time, the relative commonness of seizures in childhood—even among children who were not later diagnosed or considered epileptic—may have further motivated parents to keep seizures hidden. Whether one or more followed a fall or a particularly high fever, it likely remained the practice of many families and the advice of physicians to wait to see if a child would "outgrow" their convulsions, as some tended to do.[46]

Such considerations meant that epilepsy and seizures themselves could be undisclosed and unacknowledged even within families. It was later revealed, for instance, that Emilie Dionne, one of the famous quintuplets of the Depression-era tourist attraction Quintland, experienced seizures from childhood.[47] Along with her four identical sisters, Emilie survived a premature birth in a northern Ontario, Canada, farming town only to spend her early life under the continuous watch of government-appointed doctors, researchers, and a rapt international audience. In 1954, two years after she and her sisters were emancipated as wards of the state, Emilie died at the age of twenty following a seizure at a Quebec convent, where she was preparing to become a nun. In a manner fitting the voracious interest surrounding her family, US newspapers trumpeted "Autopsy Reveals Family's Secret."[48] Despite the exceptional scrutiny under which she led her life, her epilepsy had been news to the public and, according to later accounts, to some of her sisters as well. Having grown up highly sheltered, the surviving Dionnes had not known the word for their sister's seizures yet tacitly understood never to mention them to anyone.[49]

Similarly, American neurologist and neurosurgeon Wilder Penfield realized only later in life that his older sister, Ruth, experienced seizures. As his memoir and biographers explain, he grew up in a turn-of-the-century Protestant household where closed doors and explanations of "nerves" had shielded him from a fact that he perhaps always knew but that went unmentioned within his family. These largely unstated facts nevertheless influenced the direction of his medical career. In 1928, following his move from New York to Canada, Ruth became one of Penfield's first surgical patients in his new practice in Montreal, where he would subsequently perform more than 1,000 operations on people with localized forms of epilepsy (see chapter 5). In Ruth's case, the discovery of a malignant brain tumor as the cause of her seizures, and not the usual "cicatrix," or scarring, on the brain that Penfield often operated on, resulted in an extensive removal of tissues from her frontal lobe. She would live for three more years beyond the operation.[50]

Motifs of secrecy and revelation within families run throughout accounts of epilepsy from the first half of the twentieth century. Insofar as epilepsy remained a heritable condition, seizure-free family members were also affected by the diagnosis. During this time, eugenics normalized the practice of identifying people's pedigree in medical marriage counseling and in certain employment and educational contexts. Though many family members of people with epilepsy, and those with epilepsy themselves, married and had children in this era, cultural perceptions of epilepsy's heritability were strong, reaching their height by the late interwar and World War II period. Medical and public discussions of epilepsy following the 1927 *Buck v. Bell* ruling, for instance, noted the "great preponderance of carriers" among seemingly healthy family members and their likelihood of "transmitting a tendency to a future generation."[51] In film and fiction, characters like Mary

Lamont "bid farewell to love" as a sibling's epilepsy became proof of one's own "danger" as a spouse.[52] Others in real life, like the Dionne sisters, were warned that knowledge of their sister's diagnosis could blunt their marriage prospects.[53] Letters between Dr. Penfield and parents similarly noted strategies for preserving anonymity, such as sending the unafflicted siblings of young patients to schools in other districts.[54] As in *Dr. Kildare's Crisis*, the real or greater tragedy was often presented as residing with the wider family, who, though they seemed untouched by epilepsy, might nevertheless be thwarted in their ability to carry out the lives they desired or that were expected of them.

During the first four decades of the twentieth century, epilepsy thus appeared to be as much a family phenomenon as something experienced by seizure-prone people. It was also a condition hidden yet visible in various ways. More unclear than its effects upon families or communities in this era, however, is what epilepsy meant for those who experienced it directly. As Douglas Lamont described his situation toward the end of *Dr. Kildare's Crisis*, "A man doesn't dare take two steps away from his house. . . . Even if he stands in his room he may be struck down like a piece of falling wood."[55] Faust, the author of those words, imagined that epilepsy and seizures kept individuals sequestered and vulnerable to exposure. Real-life examples, like those of Jean Clemens and Emilie Dionne, in some ways substantiate this view. By mid-century, however, perceptions of epilepsy as a private family matter were beginning to shift, placing seizure-prone people and the invisibilities surrounding their illnesses closer to the public and to the center of epilepsy's story.

Magnifying Epilepsy: Medicine, War, and Advocacy

Beginning around 1940, a group of physicians and interested citizens endeavored to change the way epilepsy and people with epilepsy were perceived. In a campaign that for the first time targeted stigma outright, early advocates condemned the social attitudes long fueling secrecy and centuries of discrimination.[56] New organizations like the Laymen's League Against Epilepsy (LLAE) challenged mainstream assumptions about the nature of epilepsy, including its heritability, poor prognosis, and purported links to mental illness and low intelligence. Though epilepsy advocacy would differ in key ways, these efforts also occurred in close step with campaigns to prevent and cure infectious childhood diseases such polio and the lifelong disabilities it created, beginning with the founding of the National Foundation for Infantile Paralysis in 1938.[57] In the case of epilepsy, the development of special interest groups and organizations followed rather than produced changes in its medical understanding and treatment. Nevertheless, both medicine and advocacy related to epilepsy would transform rapidly together during, and especially after, World War II.

Before these events, however, the science of epilepsy began to shift with leading collaborations between researchers centered at Harvard, Columbia, and McGill Universities during the interwar period.[58] In 1928, the Harvard Epilepsy Commission (HEC) was founded under the leadership of Harvard neurologist and epilepsy specialist Dr. William G. Lennox. Seeking to expand both disparate clinical knowledge about epilepsy and resources, the group comprised neurologists, psychiatrists, and pediatricians from across four of Boston's hospitals. Most notably, the introduction of electroencephalography, or EEG, in the mid-1930s propelled the HEC's research considerably (see chapter 2). Through electro-amplification of signals from the heads of patients, EEG revealed a new language of brain waves, making a condition that was often imperceptible outside of seizures a disorder of "cerebral dysrhythmias" that could be read instantaneously and at any time.[59] In addition to improving diagnosis and identifying the invisible wave patterns present in various kinds of epilepsy, such studies contributed to a declining view of epilepsy as a family disease. In funding applications as early as 1935, HEC members reported that epilepsy was "seldom inherited" and did not appear to be *directly* heritable.[60] Rather, of far greater importance was the electroencephalographic profile of the patient in conjunction with that of his or her spouse, thus limiting the scope of concern from entire families to the individual and couple at hand.[61]

Another breakthrough came in 1937, when Columbia University neurologists H. H. Merritt and T. J. Putnam published research on the use of Dilantin (also known as phenytoin sodium) in controlling grand mal, or convulsive, seizures (known today as tonic-clonic seizures).[62] Like vaccines and other "magic bullets" that buoyed public and professional confidence in medicine throughout this era, the development generated immediate optimism in scientific circles.[63] By 1938, the drug was manufactured and marketed by the Detroit-based pharmaceutical company Parke, Davis and Company. Representing the first in a new class of "anticonvulsant" medications, Dilantin greatly reduced the occurrence and severity of seizures. It was also comparatively well tolerated by patients. In contrast, the widely used barbiturate phenobarbital, or Luminal, synthesized in 1912, though generally efficacious, was known for its strong hypnotic effects. Dilantin in many cases also replaced the starvation and ketogenic (high-fat, extremely low-carbohydrate) diets that emerged in the 1920s in conjunction with metabolic studies of epilepsy.[64] In addition to expanding the therapeutic options available for patients, Dilantin and the other anticonvulsant drugs that would follow it were vital in fostering a new language of "seizure control," which became a standard of both treatment and comportment by the early postwar era.

It was the vast social changes of World War II itself, however, that made epilepsy a more visible public issue than ever before, encouraging new types of advocacy by physicians and other citizens. As with World War I, US participation in the Second World War raised concerns about the incidence of epilepsy in the general population as those recruited and drafted underwent examination. For the first time, EEG was used to screen men for unknown or concealed epilepsies as indicated by their brain waves. Many more who had seizures in training camps or overseas were subsequently discharged.[65] Concerns about acquired or traumatic epilepsy among US servicemen also rose during this era. Anticipating the sizable social and financial cost of epileptic veterans to the American public, neurologists and neurosurgeons like Wilder Penfield emphasized the problem of long-term disabilities, like epilepsy, resulting from war-related brain injuries.[66] As his contemporary William Lennox noted in 1943, "supposedly no epileptics go into the armed forces but certainly many come out"—as many as five to fifteen of every five hundred wounded servicemen, he estimated.[67] In addition to those already in the population with epilepsy, he concluded that "a great many more men would become epileptic as a course of the war."[68]

Others participated in wartime discussions and mobilizations around epilepsy as well. In November 1940, the Laymen's League Against Epilepsy (LLAE) formed in Boston in close relationship with the HEC. While predated by other epilepsy coalitions like Europe's International League Against Epilepsy, founded in 1909, the Laymen's League brought together doctors, patients, and members of the public.[69] Apart from its medical advisory board, of which Dr. Lennox was head, the league's president and most of its members were interested donors, general physicians, people with epilepsy, and their friends and family members.[70] The latter two groups, representing roughly 80 percent of the league, contributed little monetarily beyond the annual $1 membership fee but were said to be the core of the organization.[71] Whereas the league published the names of its affiliated physicians and eminent sponsors, which by 1942 included presidents of banks and colleges as well as First Lady Eleanor Roosevelt, members with epilepsy and their families remained anonymous. In solicitations for membership, in fact, prospective members in this category were assured that their names would not appear in league newsletters, suggesting both persistent stigma and the perceived risks of exposure for seizure-prone people and their kin.[72]

The LLAE had two distinct objectives. With "feed the fires of research" featured on its letterhead, the league first aimed to raise money for medical research through a full complement of university, government, and private sponsorship. Compared with other illnesses and diseases—not least of which was polio—epilepsy, they argued, continued to be woefully underfunded. Its second but equal objective was to disseminate credible, up-to-date information about epilepsy and its treatment. In particular, the league sought to reach family doctors, who were often patients' and families' first point of contact yet could be uninformed about the most recent science or medications available. Physicians also provided a potential bridge to new members. As the LLAE's first president, Francis B. Riggs, suggested in a letter to doctors during the league's early years, "If one of your patients, or a relative, wishes to join forces with others in combatting popular ignorance and prejudice, hand him the enclosed card."[73] In its first few years of activity, the LLAE reported receiving five to six such letters from people across the country each day. Varying from handwritten notes to letters typed on business stationary, most requested advice and referrals to local specialists. They also commonly shared descriptions of the challenges that living with epilepsy presented.[74] The league responded to all letters and, as domestic war efforts grew, expanded its outreach, writing materials for the Veterans' Bureau, the Red Cross, the armed forces, and various US employment centers. By 1943, the LLAE had grown to become a national organization, with members in forty-four states and five countries beyond the United States. That year, to better emphasize its national scope, the LLAE abandoned the word "laymen" in its title and renamed itself the American Epilepsy League (AEL).[75]

The AEL, and LLAE that preceded it, built its platform and donor base primarily on the promise of scientific developments in the field of epilepsy research, like Dilantin and EEG. The suggestion of a hopeful way forward through science also allowed for a more direct means of criticizing existing social conditions. As President Riggs emphasized in the league's first newsletter, "The public is awakening to the fact that epilepsy is not inevitably incurable, and that there is an urgent necessity for further research."[76] "In a time of war," she argued the following year, "maintenance of the home front and the elimination of ignorance and sickness are essential activities which should be extended and not curtailed."[77] Materials written by Lennox also signaled these priorities. In *Epilepsy Explained*, a pamphlet distributed to members, he invoked the metaphor of "Stepchild of Medicine" to describe epilepsy's plight as a neglected and scorned illness locked "in the dark closet of public prejudice."[78] The image of medicine as an enlightening force similarly appeared in his 1941 book *Science and Seizures*, which was distributed free of charge to LLAE members. Reviews of the book by physicians and laymen cast science as the cure for popular misconceptions about epilepsy.[79] Equating the experience of reading the book with "throwing a life belt to a drowning man," one member wrote that readers would gain "through the medium of scientific knowledge a new vision of their affliction."[80] Another remarked on the book's importance to a variety of people, stating that "every doctor should have a copy in his library, likewise, ministers, social workers, educators, and criminologists."[81] Yet another suggested that greater visibility and acceptance of epilepsy might occur as a result of the publication: "A copy of the book should be kept on the table in the parlor where callers and guests will be sure to see it."[82]

In addition to naming epilepsy outright and connecting interested people across the country, LLAE/AEL bulletins and newsletters encouraged members with epilepsy to participate directly through writing. In a recurring section of the league's newsletter called "The Layman Speaks," President Riggs requested member feedback, what she described in 1942 as "crisp reports of your experiences, problems which have arisen because of seizures . . . poems, suggestions, encouragements—anything is welcome."[83] The series chronicled indignities and injustices—personal, social, economic—that Americans encountered because of their epilepsy. As one anonymously authored poem, "Home Defense Needed," expressed that year,

> We can tell you at home
> About a Concentration Camp,
> If you're fighting Epilepsy
> And Black-out like a lamp.

As we plunge all our might
In a second World War,
You still have us with you,
And we now number more.
As we liberally give
For the Nation's defense,
Remember we're sorry
To be such an expense.
From Biblical times
We've gone down in a heap;
In a wide-awake land
We continue to sleep.
I'm not proud of this handicap—
That's an honest admission,
But at least I contribute
To Harvard Epilepsy Commission.[84]

Though this poem raises questions about authorship given its reference to donations and core league arguments, "The Layman Speaks" columns also presented perspectives not printed elsewhere at the time. By underscoring the desire of those with epilepsy to participate in the war effort, for instance—or challenging broader assumptions that people with epilepsy did not "contribute"—it highlighted the marginal status of seizure-prone people through a compassionate lens. Such entries offered a rare opportunity for people to give voice to their experiences and to recognize themselves in the words of others. Though uncommon in 1942, such an approach would be modeled in a new and more public genre of writing about epilepsy after the war.

Postwar Publicity and Representation

The story of Herbert Jackson, the passenger on a New Jersey bus once turned down by the United States Army, was typical of depictions of epilepsy, and people with epilepsy, after World War II. As featured in the 1953 *Saturday Evening Post*, Jackson was the epitome of the epileptic citizen: he was honest, hardworking, a willing inductee for military service, polite and deferential to his elders, intelligent, reasonable, and "in control" of his condition.[85] Except for the exceptional circumstances of war that made him conspicuous for his age, gender, and apparent level of fitness (he wore civilian attire rather than a uniform), Jackson was otherwise unremarkable. He was thus a person with whom to empathize. That he was "invisibly handicapped," the *Post* suggested, was not disingenuous but compelling. According to new formulations, he was "near normal."[86]

The article "No Wonder Epileptics Are Bitter" in which Jackson's story appeared represented the height of a new kind of discourse about epilepsy by the 1950s. Throughout and after the war, organizations like the American Epilepsy League and American Epilepsy Society (or AES—founded in New York in 1946) became increasingly concerned with the plight of people with epilepsy.[87] Through the coordinated efforts of such groups and local and state public health leagues, epilepsy was for the first time discussed in periodicals like *Life*, *Look*, and *Ladies' Home Journal* as well as in numerous newspapers.[88] Reaching beyond the LLAE's wartime goals of educating doctors or raising funds for medical research, these articles targeted the largest possible audience. As with earlier efforts, they referenced "closeted ghosts" and "family skeletons" to underscore the need for a public unmasking of a treatable medical problem.[89] Familiar too was the presentation of medicine as a metaphorical light to be cast upon the shadows of social prejudice. Whereas epilepsy, like tuberculosis or venereal disease, was "considered not so many years ago to be a disgrace of the patient and family," modern medicine, articles informed readers, had finally made the "last hush-hush disease" a "hopeful rather than hopeless medical condition."[90]

Yet, although epilepsy and seizures were portrayed as no match for Dilantin and EEG, the social problems facing people with epilepsy were said to be more insidious.[91] Science had charged ahead, but society was deemed regrettably backward. "Much can be done for epilepsy," one 1949 article in *Today's Health* began, referencing the majority of people who experienced better seizure control as a result of Dilantin, "but what can we, the public, do about the barbaric way we react to it?"[92] Seeking to make epilepsy "a problem of public health" like any other, members of the AEL and collaborating public officials lobbied both within and outside the press. Much of their message centered on epilepsy's "social handicaps" as they advocated for basic rights and privileges in areas of work, education, and family life.[93] According to Dr. Lennox, who remained an outspoken champion of people with epilepsy after the war, marriage, children, and career were tokens of "successful living," and epilepsy persisted as a potential "handicap to success."[94] The people depicted in growing representations of epilepsy were generally young and white, ranging in age from schoolchildren to men and women in the early stages of establishing themselves. Black and Latino Americans, other ethnic minorities, older adults, and, to some degree, women typically did not feature, revealing assumptions about who most deserved redemption from prejudice and lack of opportunity.[95]

Nearly all people with epilepsy, however, continued to face exclusions and restrictions under law in this period, including marriage bans and sterilization legislation in approximately one-third of all states. There were also few protections from discrimination, including no recourse for those expelled from schools or fired from their jobs.[96] Although a 1942 study showed that 95 percent of people with epilepsy had worked in the past and were thus employable, many reported losing employment following the discovery of their diagnosis, and most of those working at the time of the survey said they kept their epilepsy concealed.[97] Men with epilepsy discharged from the armed services—a group constituting the second-largest category of discharges after those diagnosed with schizophrenia—were also not guaranteed G.I. benefits, including low-cost mortgages, subsidies for higher education, and unemployment compensation.[98]

Given these disadvantages, and the concerns about working-age men in particular, advocates in this period focused on creating new resources.[99] Following an Illinois initiative in 1945, several states launched vocational rehabilitation programs for people with epilepsy, finding manufacturing jobs for otherwise "healthy" men and placing some women in domestic and secretarial work.[100] Promoting the idea of the "self-supporting epileptic," public health pamphlets proclaimed that "9 in 10 epileptics can work!" and, moreover, that most people with epilepsy were in fact superior employees because they "overcorrected" for their disability.[101] As an avenue to employment later in life, elementary and secondary education also became a key area of concern (see chapter 3), as did driver's licensing (see chapter 4). With the help of lawyers and public officials, groups like the AES and AEL also began to question eugenics' "dead but odiferous laws."[102] Citing results from studies using EEG, they argued that with the appropriate medical counseling, people with epilepsy could most often marry and have children and should be both free and encouraged in these endeavors.[103]

Still, some of the few social studies carried out in this era show that the issues facing people with epilepsy before and during the war were slow to fade in the decades after. A set of social work theses on married World War II veterans with acquired epilepsy in Los Angeles County in 1959, for instance, found that almost half of participants were unemployed or underemployed. Only one man, in fact, had returned to the job he held prior to his military service.[104] While most participants had a tenth-grade education or higher, only one had a college degree, and only a few of those unemployed managed to obtain full disability compensation.[105] A number of those out of work cited losing a driver's license as the primary obstacle to regaining employment.[106] Though nearly all participants, to the surprise of researchers, had met and married their spouses after their injuries, fewer than half of the couples had children, and in most cases concern about transmitting epilepsy was the explanation given. Moreover, approximately half of the wives interviewed had chosen not to tell their families or friends about their husband's diagnosis, and in cases where families were informed, levels of support were mixed.[107] Researchers also found that most wives did not work, irrespective of their husband's employment status.[108] Nearly all participants, however, regardless of their race, class, ethnicity, occupational status, or religious background, reported fears of financial insecurity, social rejection, injuries at work and home, and, among couples who had children, development of seizures in their children.[109]

Notwithstanding accounts in academic studies, however, human interest stories and vignettes in the popular press remained the primary way in which the public learned about epilepsy in this period. Real-life portraits of people with epilepsy, or at least those presented as such, were effective in presenting a reformed vision of those with epilepsy as sympathetic characters, deserving of the same rights and freedoms as others.[110] In some ways, these sources traded older tropes of disability for newer ones. Representations of the obstinate, lazy, untrustworthy, and emotionally flat epileptic of the prewar period vanished—as for the most part did fictional accounts of families and individuals whose lives were ruined by seizures. In postwar newspapers and magazines, those with epilepsy were described as people who exercised self-control and sought to live and work "independently." Like profiles of "Epileptics of Note and Worth" that became popular during the postwar era and included famous figures such as Julius Caesar and Fyodor Dostoevsky, new depictions highlighted the ability, talent, and ultimately the humanity of those they featured.[111]

At the same time, postwar articles emphasized the tenuous footing that seizure-prone people experienced in their daily lives. Though shown to be "managing"—even passing as "invisibly handicapped"—those presented were usually secret sufferers, limited in less obvious ways by their epilepsy. Frequently dramatized were dilemmas of disclosure challenging masculinity in particular. In articles with titles like "Should He Tell?" readers were asked to consider the risks associated with revealing one's epilepsy to a loved one or employer.[112] A 1955 article in *Today's Health*, for instance, opened with the story of "Ted and Jane," a young couple who lived in a state that allowed them to marry. Nonetheless, Ted was, as the article presented it, "living in a personal hell . . . dreading that his employer would somehow learn, that he would be dismissed . . . and that he would be unable to support a wife and might instead become a burden to one."[113] Other articles, similar to the story of Herbert Jackson, focused on the moment of exposure and fallout. Scenarios such as "Janey's crying again . . . Janey can't go to school . . . why? Janey has epilepsy" served as common illustrations.[114] Also typical were once esteemed men shunned by their peers at work or in war, dejected schoolchildren, and, less often, women crestfallen over their abandoned plans for marriage and motherhood.[115] In articles like "Social Outcast—Age 9" and "No Wonder Epileptics Are Bitter," readers were led to see the injustice afflicting the seizure-prone person, who, apart from seizures, was "just like everyone else."[116]

The Near Normal Paradox

The connections postwar articles drew between readers and the people they featured fostered a developing notion of "near normalcy" in this era. According to such formulations, those with epilepsy were not "abnormal" as perhaps presumed, but basically, or "almost," normal. Throughout the mid-century period, concerns with what was "average" and "normal" found increasing expression in medicine, psychology, sexuality, politics, and child-rearing, among other areas.[117] Though not always easy to define, such concepts provided Americans with new ways to classify themselves and others relative to a larger, imagined public.

Articulations of epilepsy's near normalcy first appeared in the 1930s in response to eugenic estimations of low intelligence among people with epilepsy. According to Dr. Lennox in 1937, the majority of those with epilepsy were "mentally normal or near normal"; the "imbeciles and idiots, the monsters, ugly mistakes of nature" described by biologists like Raymond Pearl were the exception.[118] Whereas such individuals might require, as Lennox put it, the epileptic colony's "refuge from a cruel world," 90 percent of people, who had "normal" and "near normal" intelligence, offered what he deemed ample "social usefulness."[119] By World War II, the idea of near normalcy also encompassed perceptions of epilepsy's prevalence. As LLAE and AEL publications often cited, approximately 1 percent of the population had epilepsy—a figure that was more significant in absolute terms. In a 1945 article in *Women's Home Companion,* for instance, then AEL president, Brooks Potter, recounted feeling at first "ashamed of her affliction" but "less lonely when she read that at least half a million persons in the United States were fellow sufferers."[120] This figure was "as many as had diabetes or active tuberculosis," making epilepsy a condition both commonplace and worthy of discussion.[121]

Rapidly departing from an initial focus on measured intelligence or statistical occurrence, however, near normalcy in the postwar era came to describe the invisibility of epilepsy as well. In this framework, people with epilepsy were only "partially" or "temporarily" disabled rather than sick. New and frequently appearing variants of the term "near normal" included "no different but at time of seizure," "part-time handicap," and "invisible handicap."[122] Each described epilepsy as an infrequent episode—a seizure, essentially—and medical management as the neutralizing or normalizing force.

By the mid-1950s, seizure control was typically defined as a seizure reduction of 50 percent or more, one that occurred in upward of 80 percent of patients following the uptake of Dilantin and other anticonvulsant drugs.[123] It was not unusual for campaigners to suggest that such medications made people with epilepsy essentially or mostly "normal," if not "invisible." Likewise, with the success of EEG, epilepsy was described as a passing tendency of abnormal brain waves in this era (see chapter 2). Postwar advocates, in fact, referred to seizures not as evidence of a pervasive disorder or disease but as a standalone action in which the brain produced a rare discharge of electricity.[124] Some physicians even suggested changing epilepsy's name to "cerebral dysrhythmia" during this time, since, as articles noted, it "calls attention to the disordered brain wave pattern of an epileptic" and "won't remind people of something distasteful."[125] In new medical and public vernaculars, epilepsy became less constitutive of the individual and their identity; like seizures, epilepsy was something fleeting, anomalous, and indeed hardly apparent.

"Near normal," however, also implied a gap between the "nearly" normal and the ostensibly "truly" normal. It suggested a disconnect, however small, between outward appearances and some unspoken reality.[126] As neurologist Tracy Putnam remarked in a 1943 guide for patients and their families, the latency of seizures meant that those with epilepsy were generally indistinguishable from those without. "Seizures themselves produce no permanent changes in the patient's appearance or bodily health," he wrote. "With rare exceptions he would pass unnoticed in the throng of mankind."[127] This differed from depictions even a decade or two earlier, in which epilepsy was presumed to leave multiple—albeit, often subtle—physical and behavioral marks on the individual, ranging from one's coloring to speech patterns.[128] According to these earlier understandings, "passing unnoticed" indicated a deceit of both appearance and intent, as was shown by Douglas Lamont in *Dr. Kildare's Crisis* and the aberrant identity he harbored. By contrast, examples by the postwar period largely denied any such contradiction within the individual or else pointed to society's failing in continuing to stigmatize epilepsy in ways that made people hide. Brooks Potter, for instance, described in one article as "the wife of a socially prominent Boston lawyer," recounted being initially "alarmed" that a pamphlet on epilepsy should "reach *her* house in one of Boston's loveliest and most exclusive suburbs."[129] Her remarks echoed what Lennox had articulated about epilepsy's lacking respectability as early as the 1930s: that the "convulsions of the town epileptic are known to all . . . but the occasional seizure of the teacher, grocer, or doctor is a closely guarded secret."[130]

Indeed, epilepsy's postwar advocates increasingly reasoned that the "invisible" nature of epilepsy—including its medical control and the upstanding profile of several Americans who had it—should be a source of reassurance rather than alarm.[131] Unlike hypothetical persons whose seizures were visible and thus "known to all," those who hid in plain sight with "occasional" seizures led lives not unlike those of the imagined readers of postwar articles. A 1949 piece in *Collier's* magazine, for instance, began, "Dixie Lou is a tall, brown-eyed, nice-looking girl of nineteen who seems in the pink of health. . . . Yet Dixie Lou . . . has had epilepsy since she was twelve."[132] The rest of the story, which assured readers of the girl's robust constitution, engagement in student life, and plans for college, made her epilepsy but a small detail in an otherwise long list of admirable qualities. What "seemed to be" was therefore legitimized as real; to "pass unnoticed," as Dixie Lou might, and as Herbert Jackson did until he was questioned on a bus, was evidence of one's capacity to belong, and thus of belonging itself.

The formulation of "normal but at time of seizure" nevertheless left a sizeable caveat: the seizure itself. If seizures removed a person from the norm, then any seizure, however infrequent, was implicitly discouraged. Not until the 1960s, in fact, were details regarding different kinds of seizures, or instructions on what to do in the event of a seizure, discussed in public coverage of epilepsy. Although manuals for physicians and patients from the 1940s onward described them as medically safe occurrences with an easily manageable protocol, the public was not similarly asked to understand, support, or even tolerate seizures.[133] Just as a 1927 source reasoned that "a convulsion, because of its alarming effect on the bystanders early made an impression on mankind," so too did postwar sources for people with epilepsy emphasize the undue strain that seizures placed upon one's community and family.[134] In Putnam's 1943 guide, for instance, he argued that the young and impressionable in particular must be "judiciously prevented" from witnessing convulsions. "Children," he stated, "should be protected from seeing strong grief or fear, or of having their sympathies greatly aroused toward situations which they are powerless to help."[135] A Gallup Poll in 1949 corroborated this perspective, revealing that although 92 percent of people had heard of epilepsy, and even half knew of someone with the condition, only 57 percent of respondents would allow their children to play with one who had seizures.[136] Such losses of bodily control and personal composure away from home were not only inconvenient for others but also taboo.

The idea that seizures were private affairs, more appropriately unseen and undiscussed, persisted throughout the postwar era. At the same time, depictions of seizures as frightening and mystifying episodes continued to stigmatize people with epilepsy. Much like the unsightly drama of uncontrolled seizures presented in the final chapters of *Dr. Kildare's Crisis*, a postwar pharmaceutical ad that circulated in popular and medical sources began: "The cry, the fall, the clamping teeth, the clonic and tonic contractures, the incontinence—all may yield to Dilantin." With medical treatment, the advertisement concluded, "The epileptic may be spared his terrifying episodes and gain opportunity to lead a more normal and useful life."[137] Although such ads spoke of "sparing" people with epilepsy and allowing them "better" lives, what they equally emphasized was the response that seizures provoked in others. The "terror" most keenly implied, in fact, seemed to be on the part of the witness in hearing the cry and watching the fall. Reflecting fears and aversions like those in Putnam's guide and the 1949 Gallup Poll, the true horror of epilepsy and seizures belonged to the imagined observers of the people directly experiencing them. For those with epilepsy, the difficulty was in being observed, hence the need to prevent seizures from occurring or being witnessed in the first place.

Seizures, the new focus of beliefs once ascribed to an entire illness, marked the breakdown or "failure" of physical composure and self-control. Insofar as seizures remained a real possibility among even the best medically controlled cases, all individuals with epilepsy also continued to be vulnerable to what they represented. "Near normal," in other words, was an insufficient state for those who lived with incompatible realities of partial tolerance and the looming risks of exposure that seizures posed. Seizure-prone Americans themselves described these tensions. In 1948, following US and international media coverage of Wilder Penfield's surgeries in Montreal, dozens of people wrote to say that social uncertainty was the most debilitating part of their epilepsy.[138] One man, dissatisfied with the partial control that Dilantin provided and the double life he felt he led, explained, "I am an epileptic but this fact is <u>not</u> <u>well</u> <u>known</u>. . . . If necessary, I would spend my whole vacation in the hospital if it meant being cured" (emphasis in original).[139] Another woman described the impact that fear had on her work and personal life: "I am 22 yrs old now and I am very much in love with a very good man who wants to marry me but I have refused." "Is there any chance," she continued, "that I could ever be cured of this so that I would not be afraid to go anywhere, for fear of having an attack? . . . I've gotten so I worry all the time for fear I will have one at the office or away from home." She signed off this way: "Give me some hope of becoming a normal person again."[140] A hotel clerk described the nerve-racking nature of seizure latency: "I am just living in fear, I'm afraid to walk the streets alone, in case I take a spell, I'm afraid to eat certain things, afraid of this, that and so on, just as I said, 'Living in Fear.'"[141] Like the fictional Douglas Lamont's assessment that one dare not take more than a few steps in public, real-life accounts expressed similar desires for control as a remedy for these anxieties.

Constructing Invisibility

A common lament among epilepsy's postwar campaigners was that social prejudice forced seizure-prone people to hide their conditions in ways that evoked a harsher past. Yet in many ways, the cultural current and the goal of advocacy work after World War II was to more effectively allow those with epilepsy to do just that; rather than creating an openness around epilepsy or a tolerance and understanding of seizures, new public discourses raised awareness about the condition while simultaneously obscuring it. Near normalcy, too, was a "closet" of a kind. People with epilepsy nevertheless continued to occupy various social roles and spaces throughout this period, as they perhaps had always done.[142] Whether fictional or real, those like Herbert Jackson entered certain contexts with apprehension and dealt with consequences when their epilepsy became known. A new genre of writing about epilepsy in the postwar era recognized and even sympathized with this fact—hence the figure of the secret sufferer and his threat of exposure. However, in giving the appearance of unmasking epilepsy, as so many popular articles did, there was also the suggestion that something revolutionary was underway. Depicted in such sources was the promise of greater inclusion and a transparency around epilepsy rather than a proliferation of its polite fictions.

If seizures and their stigma persisted in the postwar era, what emerged in this period was a medical and moral emphasis on control itself. "The first duty of every person subject to seizures," Dr. Putnam wrote during World War II, "is to learn to control them. The second is to live as normal a life as possible. Every patient who does well for himself helps to dispel false ideas about the condition."[143] As seizure control and social duty became increasingly entwined in this era, they further revealed the value that society in fact placed on having an "invisible handicap." People with epilepsy were expected to keep both themselves and their seizures regulated. Those who had seizures during this period were often criticized for not alerting others to the auras that sometimes preceded their seizures.[144] As other chapters show, seizures were also presented as signs of medical noncompliance and personal mismanagement.[145] Ultimately, public support and acceptance of epilepsy relied on keeping its most visible traces out of sight—on ensuring that epilepsy was the latent and respectable condition that experts and new sources claimed it could be.

Epilepsy's coming out in a sense then was also its containment; rather than the undoing of a family and public "secret," there was instead a new organization of secrecy in the form of an invisible illness and disability. This chapter began by arguing that emerging depictions of epilepsy as "invisible" in this period took on mostly positive connotations: for some people, its presumed inconspicuousness beyond seizures became an argument for social belonging. Yet as the emphasis on genetic inheritance diminished and on medical self-management grew, epilepsy's invisibility also became an individual burden for some. Even today this feature remains a hallmark of many invisible illnesses and disabilities for which people note a lack of acknowledgment of their experiences and thus feel pressure to further keep them to themselves. The "subtle," "intermittent," or "high-functioning" nature of one's differences creates an experience distinct from other illnesses or disabilities. This distinction is true of some mental illnesses; traumatic brain injuries; chronic pain; autoimmune, sleep, and fatigue disorders; postviral syndromes; and numerous other examples.[146] For epilepsy, its history since the mid–twentieth century centers primarily on seizures and what they have continued to mean as other elements of the experience have changed around them.

Technology and Transparency

EEG Retraces the Secret Disease

Reclining on a cot in an office at Massachusetts General Hospital in 1951, electrodes stuck to his head, Albert Einstein considered familiar problems of relativity. Neurologists, meanwhile, examined the inky waves that emerged on a paper printout from a device called the "electroencephalograph" or "electric brain writer." Einstein's brain was writing those waves, and given his supreme intelligence and somewhat impenetrable theories, people were deeply interested in what his brain, by way of the machine, had to say. According to several of the next day's newspapers, the electroencephalograph provided a view of the invisible work that happened beneath the barrier of any human skull. Through immense radio amplification, electrical potentials emanating from the head could now record the lines that laid bare the physical basis of psychic function. In a short time, the technique had shed light on several disorders of the brain, could identify both family resemblances and individual differences to rival a fingerprint, and, in Einstein's case, even provided clues about the higher processes of the mind itself. The electroencephalograph, papers predicted, might soon be "spotting geniuses," too.[1]

Seventeen years before scientists sought to capture Einstein's best thoughts, however, electroencephalography and its corresponding test, the electroencephalogram—both called "the EEG"—had first established a place in neighboring labs at Harvard University specializing in the clinical study of epilepsy. Although Einstein in 1951 was not suspected of having seizures—nor was he among those notable geniuses, like Socrates, for whom epilepsy was rumored to provide the definitive "spark"—his experience was nevertheless connected to this history.[2] In fact, it was through the coordinated work of physicians and scientists affiliated with the Harvard Epilepsy Commission in the 1930s that the technology first gained its extraordinary powers to "read" and thus "write" the brain.[3] By extending expert view to uncharted depths, EEG became a trusted medium for detecting and diagnosing a once enigmatic and veiled illness—for spotting not recumbent geniuses but rather what neurologists described as both the "fugitives" from and "victims" of epilepsy.[4]

This chapter examines how electroencephalography, a technology of enhanced sight, emerged as the preferred way of knowing epilepsy—the paradigmatic "hidden" or "secret disease"—during the middle decades of the twentieth century. Though EEG had been developed decades earlier in Germany, its application in American epileptology launched the tool as a supreme authority on the brain and the object of considerable expert and popular interest. Organization around this particular use occurred precisely because EEG intervened so persuasively at the point of epilepsy's inscrutability.[5] Through unparalleled graphical evidence of distinguishable, everpresent "dysrhythmic" brain wave patterns in seizure-prone people, EEG not only created a language for reading, and indeed articulating, epilepsy—it also made the condition visible in those who were not having seizures. With the proper equipment and expertise, an inner tendency to seizures became theoretically detectable at any juncture, displacing epilepsy's most outward manifestations as the primary window on the condition.

The consequences of this moment of seemingly sudden visibility were multiple. For one, electroencephalography was of diagnostic and research value to a new generation of neurologists, who claimed the technique gave them an "explanation of an old disease" and turned "superstition into science."[6] By making epilepsy eminently readable, EEG also allowed such specialists to produce new and verifiable knowledge on the subject, thereby advancing their professional standing. If epilepsy positioned EEG machines as comprehensive writing devices and neurologists as their readers, the technology also altered the ways that people were deemed "epileptic." In addition to showing that the illness was a distinct neurological condition rather than a nebulous mental disease, electroencephalographic study redefined epilepsy as a fundamental expression of one's brain waves.[7] Like many diagnostic technologies, EEG shifted modes of investigation and evidence from broader clinical observations and patient input to the physician-administered evaluation of a test.

At the same time, electroencephalography also challenged the parameters of epilepsy itself. In addition to establishing new understandings of the various wave types found in different kinds of epilepsy, physicians soon discovered that while epilepsy was not directly heritable as long supposed, an unforeseen number of people were "asymptomatic carriers" of dysrhythmic waves.[8] Such subclinical cases complicated notions of who was epileptic as much as they stirred existing anxieties about the hidden and genetic nature of the condition. Moreover, in its apparent ability to detect and write a growing number of dysrhythmias in outwardly "normal" people, EEG also seemed capable of exposing those who attempted to pass as "non-epileptic." By revealing brain waves in contexts where epilepsy's concealment was of greatest concern, EEG allowed experts to perceive ambiguous and otherwise "masked cases."[9] In this way, the technology was for a time considered to be a suitable arbiter in matters ranging from public safety and military fitness to more intimate affairs of marriage and family planning.

This chapter advances scholarship that broadly considers the technologies experts have developed to better interpret opaque bodies. This includes those that facilitate detection and surveillance and ultimately bolster perceptions that something essential of a person—including a neurological "disorder" or "dysrhythmia"—can be plainly seen. The example of epilepsy and EEG in the middle of the twentieth century contributes to two key perspectives. First, it illustrates how illness and technologies of this kind are mutually defined and co-evolving. Second, it highlights the conflicting desires and doubts enfolded within technology's sensing power—especially as it bypasses those tested to reveal ostensible truths about human bodies and brains. Within histories of science and medicine, research on imaging devices and blood, skin, and genetic tests has demonstrated how sensing and screening technologies reprioritize the collection and interpretation of bodily information. Keith Wailoo and Robert Aronowitz, for example, show that disease-specific blood tests and mammography procedures of the twentieth century altered the boundaries of their associated illnesses, including *who* is considered sick and *when*.[10] Historians of technology and public health draw similar conclusions about technologies designed to detect lies and those created to catch even asymptomatic cases of infectious disease. Each, they argue, has expanded existing forms of social labeling and control.[11]

Histories of EEG so far, however, focus only partially on these features. As a tool of early sleep science and an instrument against which the adaptive mechanisms of the cybernetic brain was evaluated mid-century, EEG has been comparatively stable and sometimes even secondary in relationship to what it has measured.[12] By focusing instead on the connections between epilepsy and electroencephalography—by placing EEG on equal footing with epilepsy, the subject it first coherently "captured"—this chapter seeks to understand how the two were formed together and what those relationships reveal.

Drawing on studies of analogous technologies and tracing EEG through multiple phases, the chapter examines how electroencephalography for a time both eased and stoked anxieties about epilepsy as an "invisible" condition. While initially addressing doctors' desires to better understand and perceive the disorder, EEG by the 1950s had complicated the picture of both epilepsy and the technique itself. Growing ambivalence regarding the powers of the technology to see fully or clearly occurred for several reasons but ultimately reflected broader uncertainties about making epilepsy more visible. In particular, efforts to identify people beyond clinical contexts were increasingly at odds with postwar values of integrating those with epilepsy and viewing them as "normal."[13] Moreover, though the technology fell short of providing the extended sight scientists once imagined it might, electro-encephalographic renderings of epilepsy by the 1950s did more to enmesh than separate the condition from its associations with criminality, genius, and sexuality. Indeed, rather than brain waves tracing the new story of epilepsy experts claimed it would, EEG in some ways more closely aligned "superstition" with science.[14]

Developing a Seeing Instrument

According to American electroencephalographers Frederic and Erna Gibbs, EEG had been developed in the 1920s to amend a particular human "blindness." As the husband-and-wife team estimated of the "electroencephalographer's place in nature" in volume 1 of their multivolume *Atlas of Electroencephalography*: "We are not equipped, unfortunately, with organs sensitive to electricity as our eyes are to light. If we were we could receive through our primary senses a highly informative view of our internal and external universe."[15] They continued:

> Electrical engineers have done their best to correct our blindness; they have given
> us instruments which are in effect high-speed broad focus "eyes" with which to

see fluctuating electrical fields. The scene that is thus revealed may at first seem strange, but it should become quickly familiar because it is not new; it has been there all along and it is in fact only an extension of the view we have seen for thousands of years with our limited primary equipment.[16]

According to the Gibbses, such enhancement of the electroencephalographer's own humble sensory "equipment" created a true and trustworthy picture of inner electrical phenomena imperceptible to the naked eye. Presented as at once plain scientific phenomena and near conjuring tricks, brain waves were natural, ever-present, and waiting to be discovered with the proper mode of sight. By the time of the Gibbs's third and final *Atlas*, they marveled at the technology's ability not only to "see" but to literally render the invisible visible: "Without opening the skull, without piercing the skin, with no manipulation of the patient except parting his hair and applying small discrete electrodes, a trained electroencephalographer can make visible part of the invisible realm of the submicroscopic and submolecular organization that underlies brain function."[17]

In 1934, the Gibbses were part of a team at Harvard that used and greatly improved upon electroencephalography and the EEG machine through the systematic study of epilepsy. That year, the Harvard Epilepsy Commission, founded six years earlier, had approached the pair to conduct preliminary studies with the instrument. Using a rudimentary device, Gibbs and Gibbs began recording action currents from the brains of laboratory animals in which they induced convulsions by means of electric shock. By comparing resting and convulsive states in the animals, the Gibbses soon "perfected their modest technique and apparatus," gaining key insight into the wave patterns of seizures and the inhibition of nerve activity that preceded them.[18] By the following year, the Gibbses and the president of the HEC, William Lennox, published their first of many papers on their electroencephalographic findings, noting the transformative value that EEG would have for the field of human epilepsies. Although earlier metabolic and endocrinological work had already established the HEC as an international leader in epilepsy research, members would later credit EEG with galvanizing this position: it was soon dubbed the "new technique" by which "the whole field of research in epilepsy moved ahead rapidly."[19]

Electroencephalography, however, was neither invented at Harvard nor developed to understand and diagnose epilepsy. In 1924, German physiologist and chief psychiatrist at the University of Jena, Hans Berger, first built a device that amplified the brain's electrical activity from the surface of the head.[20] Although the observation of electrical phenomena in the brain dated to the final quarter of the nineteenth century, when several investigators demonstrated motor excitability in the cortices of living animals, Berger was the first to confirm its presence in humans by attaching small electrodes to the head and wiring them to a modified radio receiver.[21] Using far greater amplification, the resulting "electro-encephalograph" was much like the electrocardiograph, used since the turn of the twentieth century to register and visually represent electrical currents emanating from the heart.[22] By 1929, Berger's findings established him as the "founder of psychophysiology." Nevertheless, a lack of clarity and consensus remained about what the "rhythms" of his device revealed.[23] Berger's earliest attempts to correlate brain waves with psychiatric phenomena, for instance, produced few conclusive results, while his limited observations of epilepsy in 1932 failed to show the characteristic spike-and-wave pattern that the Gibbses and Lennox would identify.[24] EEG's primary use was therefore largely unassigned during its first decade of existence, and the seeming "code" of brain waves was mostly indistinct.[25]

In hindsight, members of the HEC would deem epilepsy the most "amenable" condition to an interpretation of brain waves and thus the "chief beneficiary" of the technology.[26] In the early 1930s, however, physicians affiliated with the HEC, including Dr. Lennox and his colleague psychiatrist Stanley Cobb, recognized only a faint opportunity in EEG to reveal vital information about epilepsy and seizures. Despite Harvard's central role in epilepsy research since at least the founding of the HEC, the field had remained somewhat stagnant. Members attributed this partly to social stigma but mainly to medical perceptions of epilepsy as an "opaque" condition for which there was little hope of understanding the organic causes. Epilepsy was notoriously difficult to discern and treat, and the HEC had initially cast their focus throughout the body, studying the hormonal and cerebral circulatory systems as well as the effects of dietary and starvation therapies. As the committee reported in the *Harvard Medical Alumni Bulletin* in 1932, their progress was stalled primarily by the fact that "the brain has proven a most difficult organ to study."[27]

The Commission's earliest findings with EEG, however, suggested new possibilities. With successful applications to the Macy and Millbank Foundations in 1935, the HEC acquired the necessary funding to build an EEG instrument capable of making recordings in humans.[28] To construct it, the group employed Albert Grass, a recent graduate and instrument engineer at MIT and later proprietor of the Grass Instrument Company. Working at home with the help of his brother, Grass constructed a three-channel EEG machine for the HEC in a matter of days.[29] The machine facilitated observations in epilepsy patients at Massachusetts General and Boston Children's Hospitals, showing both the "fingerprint" uniqueness of brain waves and their common patterns. By 1936, Lennox extended his proposed studies to 1,500 patients—an ambition that quickly ran up against the immobility of Grass's hefty machine.[30] Petitioning the university for further funding that year, Lennox explained that "for obvious reasons, it is out of the question to bring all of the cases we wish to study to the lab. The program calls for a portable apparatus."[31]

The subsequent Grass Model I, which could be carried between hospitals and connected to multiple research subjects in a time-efficient manner, greatly expanded the Harvard team's collection of brain waves.[32] Within only a few years, Lennox and the Gibbses had taken several thousand electroencephalographic readings in a wide range of patients with histories of seizures. Their samples were more than large enough to distinguish between the "abnormal" patterns, or "dysrhythmias," of several types of epilepsy. "Petit mal" seizure sufferers, for example, showed alternating slow wave and fast spikes, whereas those with "grand mal" seizures were characterized by rapid high peaks and valleys.[33] Lennox and Gibbs's data also included family members and control groups, which helped define the broader variation of human brain rhythms. Each of these came to fill the many pages of the Gibbs's definitive three-volume *Atlas*. By the 1940s, not only did epilepsy have a discernible appearance that could be amplified and outlined but it was also written in the most literal sense, set in print, and published for expert reading.

Wave Diffusion and Applications

While the EEG machine ensured the electroencephalographer's "place" and Harvard's position as the "center in work on epilepsy," it also found use as a "seeing instrument" in several clinical and extra-clinical contexts beyond Boston-based labs and hospitals.[34] The high cost of the machine, maintenance demands, and the need for a trained person to administer and evaluate the test did, however, slow early dissemination. Nevertheless, by the early 1940s, Albert and his wife, Ellen Grass, had begun to manufacture commercial instruments to keep pace with growing demand.[35] In addition to indexing and defining epilepsy, EEG also fostered the development of new therapies. Given the technology's capacity to trace the "visible actions" of seizures and instantaneously "read" the brain's response to various stimuli, one if its earliest uses was to examine the effects of several existing medications.[36] Those same features also laid the groundwork for a new class of anticonvulsant drugs, beginning with Dilantin in 1938. In such studies, records of the brain's activity during and following seizures—first in animal subjects and then humans—were crucial to establishing new research methods, including ways to "objectively record" therapeutic responses to experimental medications.[37]

As a reputed "window into the skull," EEG was also influential in advancing surgical treatments for epilepsy underway at the Montreal Neurological Institute.[38] In 1937, Herbert Jasper, then a psychiatry fellow at Brown University, packed an EEG unit he built into the trunk of his car and drove to Montreal to meet the MNI's chief neurosurgeon, Wilder Penfield.[39] Having learned of Penfield's technique of electrically stimulating the exposed cortices of surgical epilepsy patients to pinpoint the area of the brain responsible for seizure activity (see chapter 5), Jasper hoped to use the device to better locate seizure foci. Following a yearlong trial during which Jasper commuted back and forth across the border with his EEG machine in tow, Penfield offered him a permanent position.[40] By 1940, Jasper had his own department in the Institute's basement and was making thousands of recordings each year. Greatly aiding in the preoperative evaluation of patients, electroencephalographic examination became one of three criteria for surgery, eliminating those patients who only "appeared" to be good candidates. By 1941, Jasper's work also came to include EEGs taken directly from the cortices of awake patients during surgery.[41] Rather than a collector or "cartographer" like Penfield or the Gibbses, Jasper regarded himself as a "voyager" and "captain on a sea of brainwaves."[42] Suggesting new scientific terrain, his characterization highlighted the novel vistas and opportunities for discovery that EEG and epilepsy together afforded.

The promise of discovery also resonated with the public as electroencephalography found its place in popular media.[43] As early as 1937, newspaper headlines such as "Man's Mind Traced by His Electricity: Also Tests Identity" and "Active Index of the Brain" spurred broader interest in the "electric brain writer."[44] Largely ignoring the technology's use in clinical epileptology, such coverage emphasized EEG's potential to recognize something definitive of human individuality and identity—as one article suggested a "new picture of electrical man."[45] Unique in its power to perceive invisible bioelectrical currents in the "most human" of organs, electroencephalography also seemed capable of solving various "mysteries" of the day, including whether or not the ever-popular Dionne quintuplets were "truly identical."[46] It was likened to X-ray vision, a personality test, a mind reader and thus "truth" teller, and a transcriber of thoughts and feelings worthy of science fiction. Lennox acknowledged such reports as "good advertisement" for the work of neurologists but also complained of sensationalism, arguing that public demonstrations performed by colleagues undermined real research. As he wrote to the president of Harvard that year: "I am getting letters from all over the country which is a time consuming nuisance, but which indicates to me a reservoir of interest when something really important like the electroencephalograph is made public."[47]

Others interested in epilepsy, however, saw value in EEG's wider exposure and what it could contribute to changing perceptions of the illness. Early advocates capitalized on public curiosity surrounding the device itself. By the early 1940s, a variety of publications for physicians, patients, and their families discussed EEG machines and their output in an increasingly approachable manner. As one pamphlet explained, "Every person has his own brain wave pattern just as he has his own handwriting."[48] Like writing by hand, undergoing an electroencephalogram was cast as a "simple and painless" process.[49] Describing the ease and even comfort of the procedure, the pamphlet continued: "The person sits in an easy chair and a number of small flat discs, called electrodes, are attached to various parts of his head by dabs of collodion. Fine wire leads from these discs through a series of radiolike boxes, which amplify the currents. The pens record the brain currents as a series of little squiggles or waves."[50] In such representations, the familiarity of easy chairs, radios, and pens, as well as the casual talk of dabs and squiggles, demystified and even humanized the technique.

EEG permeated several less expected contexts as well. In 1943, before many hospitals in the country possessed the device, the White Special School at Kathleen B. White Elementary in Detroit purchased an EEG machine from Rahm Instruments with the help of the city's Board of Education (see chapter 3).[51] The school, founded only one year before EEG's integration into American epilepsy research, relied heavily on the technique from its early

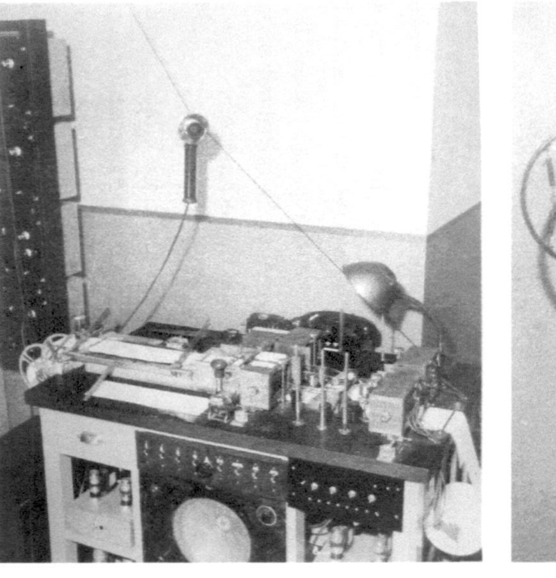 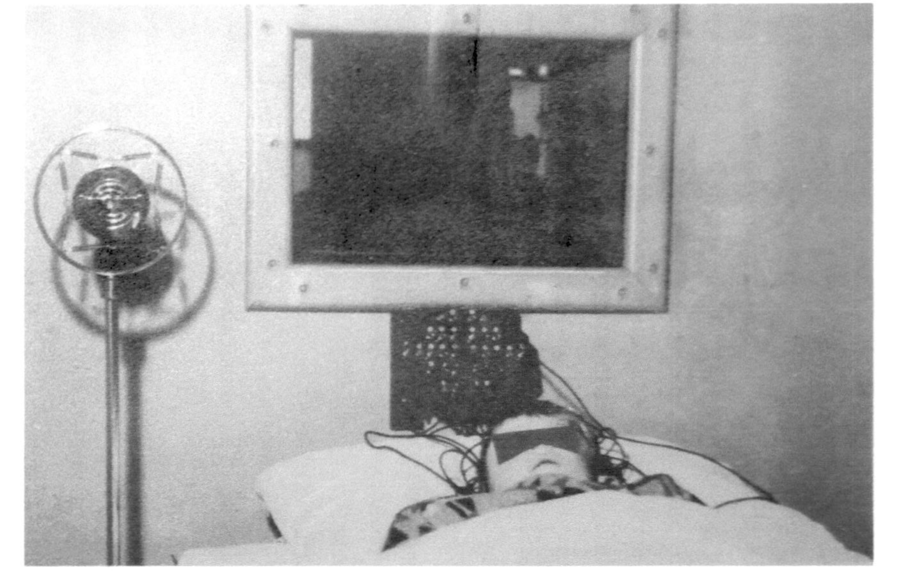

Electroencephalograph, as shown in a 1940 magazine. Original caption: "It resembles the older and more familiar electro-cardiograph" using "tubes similar to those in a radio set." From *New World Illustrated*, May 1940. Image courtesy of the Osler Library of the History of Medicine, McGill University years. Electroencephalography provided a method for evaluating students with epilepsy as well as for monitoring the effects of medications tested at the school in conjunction with a local pharmaceutical company.[52] In 1941, the school's director of special education, Alice Metzner, contacted Lennox and accepted an invitation to visit his labs in Boston. Modestly identifying as a "non medical person," Metzner regarded EEG as a "practical help to us in our problems" and expressed hope of moving all testing from Detroit's Harper and Henry Ford Hospitals to the school itself.[53] Wartime restrictions initially limited availability of the materials needed to build the device, but soon the school's EEG machine became an important fixture at the school and in the community.[54] Employing its own technician, the school examined every child in the city "in need of a test."[55] It also tested people experiencing dizziness and fainting after industrial injuries, as identified by the city's Department of Safety.[56] In 1948, the school even upgraded to an eight-channel model to meet its testing needs. Far from being a narrowly defined or expert-controlled device, the EEG machine was applied in diverse ways based on perceptions that electroencephalographic measures surpassed other modes of inquiry.

Rewriting Epilepsy

Apart from charting the rhythms of the brain and its multiple variations, application of the "brain-writer" to epilepsy after 1935 also redefined epilepsy itself. Reinterpreted through the language of the technology, epilepsy was increasingly understood as an expression of electrical currents and brain waves. By 1950, for instance, Lennox and the Gibbses had based their conclusions on more than 10,000 patients and 1,000 "normal" adults of varying ages, backgrounds, and states of consciousness.[57] From the reams of linear data collected in these and other tests, physicians became confident in their perception of an "archetypal" wave and its "epileptoid" divergences.[58] Unlike the "remarkably consistent" brain-wave patterns of "normal" individuals, people with epilepsy revealed patterns that were either "too fast," "too slow," or "too large." Electroencephalographers broadly dubbed these differences "cerebral dysrhythmias."[59]

Before neurologists' use of EEG, epilepsy had been identified mainly by the history and presentation of seizures. Though patients' histories continued to matter to doctors, electroencephalograms also rose in importance. Some seizure types, including grand mal seizures, were highly visible

Man with epilepsy having his brain waves read, Detroit, Michigan, ca. 1946. Photo by Mark Kauffman. Mark Kauffman / The LIFE Picture Collection / Shutterstock

to seizure-prone people and their families. Yet seizures of all kinds were intermittent and could be difficult to discern. Although seizures were long understood as manifestations of a broader "trouble" within—British neurologist John Hughlings Jackson began to associate seizures with specific regions of the brain in the 1860s—they provided a limited window on epilepsy. EEG, on the other hand, alleviated many of the challenges of clinical observation. No longer constrained by the timing of seizures, their tendency to come and go, or the capacity of patients and witnesses in accurately reporting and describing them, physicians could use EEG to "see" epilepsy with a widening lens.

This new perspective became a dominant feature of the 1952 novel *The Lovely Season* by American author Virginia Evans. Centering on a newlywed couple, Mr. and Mrs. Harrison, the novel dramatizes the often undetectable nature of epilepsy in ways similar to, but also distinct from, the 1940 novel *Dr. Kildare's Crisis*. The book is written largely from the perspective of Maggie Harrison, a young New York typist who remains unaware of her seizures for years, experiencing them only as occasional blackouts, fainting spells, and headaches. After her husband, Frank, witnesses a seizure one morning—the first he sees of her rapidly worsening condition—he sends her to a neuropsychiatrist, who orders a "brain wave test."[60] In what Evans describes as a procedure that "did not take long" but costs "half of Frank's weekly salary," Maggie receives a diagnosis that she deems shocking but irrefutable.[61] Once dismissive of the doctor's hypothesis of epilepsy, stating "all I do is faint," the young woman is entirely credulous once an electroencephalogram shows the reality of her situation.[62] Although her husband witnesses a seizure (an event he does not fully understand and cannot bring himself to name), causing her to undergo the exam in the first place, it is the test that holds diagnostic power and shapes the remainder of the plot between the young couple.

Part of EEG's allure was that it promised to be revealing regardless of the circumstances or persons involved. Sensitive enough to register the subtlest of wave disturbances in those with epilepsy, the technology prevailed even if the patient had never to their knowledge experienced a seizure.[63] As early as 1935, members of the HEC reported "little doubt that the epileptic seizure is preceded by a sort of electrical brainstorm."[64] Grand mal seizures were predictable hours in advance if the "increased frequency of abnormal waves" was properly witnessed.[65] Furthermore, neurologists believed that nearly 95 percent of patients showed "transient disturbances of rhythm," also termed "subclinical or inward seizures," which gave "some indication of the type of physical seizure the patient ordinarily has."[66] Unlike more outwardly visible but fleeting seizures, EEG identified the continuous rhythm of epilepsy, thereby diminishing the clinical importance of seizures.[67] As Frederic Gibbs later summarized: "What an observer sees as an epileptic fit . . . is but a small part of a much larger picture. The hidden part is most interesting and informative, for it is the part that tells what goes on inside the patient's brain. . . . When this is brought to view, we see that the old disease, epilepsy . . . is but the outward manifestation . . . of cerebral dysrhythmia."[68]

Dr. Lennox and his followers regularly argued that making brain waves visible had transformed "superstition" into "science." Rather than the "blind alleys" of previous research, EEG rather quickly "demystified" epilepsy, writing an "old disease" anew.[69] In *Science and Seizures,* Lennox devoted a whole chapter to the technique, entitled "Radio to the Rescue," in which he posited that Hans Berger's "breaching of the bony bastion of the skull" had "released epilepsy from the crypt of the unknown."[70] "Now, like turning on the lights in a dark, ghost-haunted room," he wrote, "a study of brainwaves has demonstrated that the malign forces which cause epileptic seizures are only familiar electric forces, analogous to the currents which operate our beloved household gadgets . . . tangible evidence that the disorder is physical and not metaphysical."[71] By revealing unseen cerebral activity, epilepsy could suddenly be read as easily as a heartbeat, and with remarkable accuracy. In providing "tangible evidence" and making the patient as comprehensible as a "household gadget," EEG, physicians imagined, could eliminate stigma as well. The technique, in short, seemed capable of liberating those with a condition previously feared as "metaphysical" and hidden beneath the "bony bastion" of the skull.

Discourses about "seeing" epilepsy were central to public education campaigns that sought to "bring epilepsy out from behind the veil of shame and secrecy" after World War II.[72] Although electroencephalography did not change the course of the illness therapeutically, as did Dilantin, groups like the American Epilepsy League increasingly referred to EEG as the guiding emblem of "hope" and purported "triumph" of epilepsy. In addition to the diagnostic powers of the technology itself, many argued that EEG had the power to author new social realities for epilepsy sufferers. As one article outlined in 1945: "The first step in teaching the public what it ought to know is to get rid of the word epilepsy and its connotation. . . . Specialists are beginning to talk of 'cerebral dysrhythmia,' a term which has come out of the study of brain activity with the electroencephalograph."[73] Others described epilepsy as a product of the brain's electrical misfiring, and seizures as the oft-cited "electrical brainstorm."[74] Although terms such as "dysrhythmia" and "storm" came with their own connotations of disease and difference, efforts to rename epilepsy foregrounded the diagnostic method, suggesting that a more technical and thus less stigmatizing explanation could be written. Images in popular articles of the period frequently featured physicians reading brain waves.[75] In accompanying text, the mechanism might be said to "show electrically what the brain is doing," or reveal that "an epileptic attack" was simply "neuronal distress" or an "outpouring of electrical currents."[76] One article described and normalized this suddenly transparent event: "Many body activities produce tiny electrical currents—the heartbeat, the moving of muscles . . . life itself may be described as a part of a collection of electrical phenomena."[77]

As epilepsy became a rational and readable bioelectrical condition, electroencephalography was the primary means by which someone was deemed to have epilepsy. Much like suggestions by historians that X-ray technology changed how physicians diagnosed broken bones after 1900, epilepsy by the 1940s increasingly became a condition confirmed and exemplified by a test.[78] In effect, as epilepsy became defined by the shape and speed of one's brain waves, diagnosis shifted from external means to a more interior-focused, technologically verified orientation. Hans Berger's own wife, Dorothy, for instance, was said to have benefited from an electroencephalogram that revealed the "epileptic" rather than the presumed "hysterical" nature of her seizures.[79] Similarly, students with epilepsy at the White Special School in Detroit were often admitted according to their EEGs rather than just their personal histories and other types of records. Children were rated on a one-to-five scale, with one and two being "perfectly normal," three "borderline," and four and five "clearly showing epilepsy." Those in the latter groups were deemed "diagnosably abnormal" even when they exhibited "no outward manifestation of seizures."[80] These metrics also became the way that many students understood their conditions, with some making reference to electricity in their brains, and many able to state to doctors and visitors to the school that they were "4s," "5s," or "borderline 3s."[81]

The shift toward interpreting epilepsy through EEG also had consequences for people who did not experience seizures or identify as epileptic. By 1940, electroencephalographic findings suggested that the brain waves of nearly 10 percent of "normal persons" exhibited "abnormalities" similar to those seen in patients with epilepsy and its so-called "related conditions."[82] Lennox and his team also found that approximately half of near relatives and almost always one parent of patients with epilepsy who "are and seem normal" were "asymptomatic dysrhythmic."[83] The "mother with migraines," Lennox explained, was a particularly frequent "unseen carrier."[84] Such examples amounted to nearly 12 percent of the population, or almost 15 million Americans, who were thus reassigned from the category of "normal" to "subclinically epileptic," "asymptomatic," or at the very least "carriers" of cerebral dysrhythmias.[85] As Lennox concluded of the disorienting projections, "these studies broaden the whole conception of epilepsy and require reconsideration of classifications which are based simply on the actions or the thoughts of the patient."[86] Unlike deceptive exteriors and the perspectives of patients, EEG was deemed the "best discrimination between healthy and diseased groups" and all of those in between.[87]

Family members and notions of heredity were particularly impacted by this changing portrait of epilepsy—both in its expansion as an illness category and in its connections with other conditions. Although early electroencephalographic studies were said to "clarify the problem of the inheritance of epilepsy," noting that it was not directly heritable in more than 85 percent of cases, EEG at once removed and more deeply implicated family members in diagnosis.[88] As Lennox and the Gibbses explained in 1939, "the opinion that heredity is an important influence in epilepsy is widespread," yet "only one epileptic person in five is able to name any relative who has been similarly afflicted."[89] Instead, they increasingly regarded brain waves as an "inborn" characteristic.[90] Like other physical features, cerebral dysrhythmia, and thus the *tendency* to seizures, could be passed on rather than being solely the result of environmental factors. As the team explained in 1940, "cortical dysrhythmia, although a transmissible constitutional characteristic, is not laid down, like a fingerprint, in fixed tissues, but is functional and physiologic, an ever shifting shadow of a disturbance of the fundamental electrochemical activity of the brain."[91] Whereas physicians since antiquity had spoken of a "family illness" and "blood" characteristics, those in the 1940s cited "inherited waves," "family resemblances," and "seizure predisposition."[92] EEG simply offered the best chance for tracing and understanding the recessive flow and generational rhythms of this more complex "shadow of a disturbance."[93]

Spotting Epileptics

Credited with alleviating the "suffering" of people with epilepsy by "bringing epilepsy to light," experts considered EEG equally capable of counteracting what they broadly perceived to be a significant concealment of the condition by the early 1940s.[94] Given the enduring social intolerance of epilepsy and seizures throughout the period—and the fact that more unknown than known cases were believed to exist—EEG seemed ready to identify "hidden" and unseen people. As the *New York Times* reported of epilepsy's surprising ubiquity in 1940, "There is reason to believe that for every person who is an evident epileptic there are twenty or twenty-five more who show no symptoms and are even unaware."[95] The preponderance of such "non-evident" epileptics, as well as the true "epileptic nature" of many blackouts and fleeting losses of memory, led the article to conclude that only "brain wave records may make it possible to identify those who may be subject to such attacks."[96] As one retrospective similarly described of the technique years later, it rather easily "disclosed" instances of "masked epilepsy."[97]

Stereotypes of concealment prevailed in medical and popular culture at this time in part due to the intermittent and often socially invisible nature of seizures. "Contrary to general public belief," one pamphlet explained in 1949, ". . . epilepsy need not run the entire gamut of violent contortion . . . culminating in a fall to the floor or pavement."[98] Rather, transient and indiscernible episodes, "not apparent to the casual observer," allowed many to move through the world undetected.[99] As another article presented it, "You see them on the streets every day—in restaurants, going to the movies, walking in the park. They look like ordinary people. And indeed they are. . . . Yet their lives are haunted by a secret."[100] Electroencephalographer Herbert Jasper used "fugitives from epilepsy" to characterize this seeming disconnect, referencing both the stigma that made individuals hide and their perpetual flight from the "horns of an epileptic dilemma."[101] Although such accounts suggested that epilepsy was nothing to be feared and in fact that people with epilepsy were wrongly discriminated against, the rhetoric of something untoward beneath the surface also remained a powerful trope. EEG in equal parts allowed neurologists to visualize dysrhythmic brain waves, with or without the "dilemma" of a seizure, and to detect those who might claim to have epilepsy but in fact did not.

In light of rising concerns that people with epilepsy and their dysrhythmias "passed unnoticed," doctors and various officials came to promote electroencephalography in a growing number of arenas.[102] Looking beyond traditional clinical spaces of the 1930s, they proposed EEG as a screening tool in contexts ranging from drivers' licensing bureaus and adoption clinics to ports of immigration and the military.[103] Notably, both the US and Canadian armed forces first used EEG machines to test their recruits for unseen disability during World War II.[104] Under selective service regulations, men with epilepsy were automatically classified as 4-F and rejected for service, a rule that had disqualified seven of every 1,000 men during the previous war.[105] Epilepsy also ranked highly among the reasons for medical discharge. While doctors generally attributed such cases to head traumas acquired during war—later interpreted by Dr. Lennox as the "sparks" that ignited "abnormal waves" in predisposed men—others with epilepsy were said to have evaded earlier detection.[106] Of particular suspicion were men with less noticeable "petit mal" seizures (seizures without convulsions) who "slipped through the cracks." Regardless of seizure type, however, electroencephalographic testing appeared to bypass the difficulties inherent in asking men to honestly and knowledgeably answer whether they had "ever experienced a seizure or blackout."[107] As Lennox explained, "When electroencephalographic examination is not available, the diagnosis of epilepsy rests not on physical examination, but on the history. If the registrant is unaware of his diagnosis or conceals it, he may pass and does, by the thousands."[108] With US military support and collaboration in 1941, Lennox was successful in instituting EEG in select army screening centers.[109] The US Air Force, as historian Kenten Kroker documents, also adopted testing that year to screen for pilots whose "epileptoid dysrhythmias" seemingly made them susceptible to mental breakdown.[110]

Beyond the military, EEG was also institutionalized as a screening device in certain educational contexts.[111] At Detroit's White Special School, for example, the technology was used to identify "epileptic potential" in the brain waves of school-age children—a demographic that seemed both at high risk for epilepsy and most promising to treat. Administrators regarded their World War II–era machine as the means to "locate potential epileptics in order to start treatment and care early," before seizures became an alleged "habit of the brain" and a social problem later in life.[112] In addition to identifying less obvious cases for inclusion in its program, school administrators used EEG to exclude cases that merely "imitated" epilepsy. Having consulted at length with Dr. Lennox on the difficulties of diagnosing "borderline cases" in 1941, Superintendent Metzner admitted that EEG also "served to weed out possible cases of wrongful admittances to the school."[113] Indeed, though several hundred children in the Detroit area underwent testing each year—902, for example, during the 1945–1946 school year—enrollment remained steady at around 177 students per incoming class.[114]

By the 1940s, EEG also provided vital evidence in a number of court cases. More specialized in its application than the roughly contemporaneous polygraph or "lie detector" test, electroencephalograms were typically used to determine or corroborate a defendant's diagnosis of epilepsy. As with criminal insanity, it became necessary to prove if a "fit of epileptic furor" or automatic, unconscious behavior of any kind caused the crime and thus whether the accused was criminally culpable. The first case of this kind was the 1943 trial of Edwin Codarre, a thirteen-year-old boy accused of murdering a nine-year-old girl in New York state. A one-time runner-up in a better baby contest, the controversially diagnosed Codarre was also the youngest person in the state ever prosecuted for first-degree murder. Although the two EEGs obtained during his trial became less relevant when he eventually pleaded guilty to second-degree murder, they nevertheless featured prominently in the debate and media coverage surrounding his case.[115] In other instances, EEG was mobilized to pierce the "epilepsy defense" of accused killers, including a Connecticut man charged with murdering his mother in a high-profile case in 1947.[116] To address what historian Ken Alder describes as society's desire for a just and impartial arbiter of human action, EEG, similar to the polygraph, was intended to measure guilt by confirming or denying diagnosis.[117]

Electroencephalography's proponents also recommended the technique in marriage and family planning contexts. Showing that eugenic beliefs evolved more than they declined in this era, early 1940s news articles reported that EEG could "improve the race biologically" by "detect[ing] carriers of hereditary epilepsy."[118] Lennox and the Gibbses challenged theories of direct heritability yet had noted the "eugenic possibilities" and "genetic value" of the test early on. In particular, Lennox's electroencephalographic projections by the late 1930s that roughly one in every eight Americans carried a seemingly transmissible predisposition to seizures led him to promote EEG's use in physician-led marriage counseling.[119] Writing in support of a primarily positive rather than negative eugenic approach, he explained that while people with epilepsy were not hopeless marriage candidates, they and their prospective or existing spouses would benefit greatly from screening. The chance of having children with seizures, he argued, was significantly reduced and almost negligible if the selected partner's EEG was "normal."[120] "Without such testing," he warned, "advice about the marriage of a person from an epileptic family" or with epilepsy themselves was "pure guesswork."[121] Electroencephalography thus increased peace of mind while promoting couples' procreative prerogative and the physician's place in guiding intimate family matters.

EEG intervenes in precisely this capacity in the postwar novel *The Lovely Season*. Following Maggie's unfavorable test results, her husband, Frank, also submits to an EEG test at the prompting of her doctor, who, in his simultaneous role as marriage counselor, reassures them that "epileptics hardly ever had epileptic children"—a risk of "forty to one."[122] With Frank's examination, however, the neuropsychiatrist regretfully informs the couple "on the basis of the two electroencephalograms" that they "could have done better physically picking out marriage mates."[123] Unlike earlier examples, such as *Dr. Kildare's Crisis* (see chapter 1), in which a family history of epilepsy condemns the couple (despite the fact that neither party has epilepsy), *The Lovely Season* presents the possibility of a viable marriage between seizure-prone and non-seizure-prone partners. Having exposed the unseen realities of Maggie and Frank's brain waves, however, the remainder of the novel progresses from the assertion that the couple is bioelectrically mismatched and should not have children. While each "alone," according to Dr. Guest, presents "no problem" reproductively, Maggie's condition combined with Frank's personal and family history of migraines nearly ensures that the couple will have electroencephalographically abnormal children and grandchildren.[124] Emphasizing that the otherwise happy and well-suited young couple is incompatible, the doctor reiterates the power of the technique to discern in such matters: "If there were such a thing as a cold scientific measure of marital partners, you would be judged unsuitable for each other."[125]

Delinquents and Visionaries

Following EEG's prolific writing of cerebral dysrhythmias in the 1930s and subsequent expansion into wider screening functions by the 1940s, investigators applied the technique to a broadening array of research topics. Post–World War II doctors and scientists used EEG to examine issues ranging from psychopathic tendencies and juvenile delinquency to human sexual response. Yet just as the "dysrhythmias" of epilepsy had formed the original patterns against which neurologists first defined "normal" brain waves in the 1930s, EEG's application beyond epileptology during and after the war drew strong implicit, and sometimes explicit, comparisons to epilepsy itself.[126] As a result, postwar electroencephalographic research reinvigorated a number of epilepsy's longer-standing associations with criminality, pathology, mysticism, and genius. Often deemed, in fact, an appropriate research tool where such associations already existed, EEG's wider use reified rather than rewrote many of the narratives about epilepsy that several physicians and advocates of the period sought to abandon.

By the early 1950s, for instance, EEG was deemed a promising technique for understanding criminal and antisocial behavior and, in particular, the seemingly growing problem of juvenile delinquency. Reporting on the issue of "delinquents' brains" in 1954, the *New York Times* explained: "If Jimmie joins a gang, becomes a holy terror in the neighborhood . . . and clashes with the police, psychiatrists and welfare workers are often successful in tracing the cause to a miserable home life, the conditions that prevail in the slum where he lives and plays, or the insecurity that is the consequence of race prejudice. Now doctors . . . have evidence that something may be wrong with certain areas of Jimmie's brain."[127] Separating diffuse social and environmental factors from the hard science of the electroencephalogram, the article went on to state that the brain waves of "bad boys and bad girls" were clinically "abnormal."[128] Specifically, those same children who "make life miserable for teachers and plague the police" produced readings that "closely resemble those recorded in epilepsy."[129] As Lennox previously hinted of the cerebral activity found in epilepsy, now seemingly applicable to "problem" kids, "It is as though instruments of an orchestra began to play in unison instead of in harmony, or as though representative government gave way to a totalitarian state."[130] These children, despite being as "electrically wrong" as those with epilepsy, could fortunately be identified with EEG before their troubles began. Anticonvulsant medications, moreover, might ultimately help "keep Jimmie in step with society."[131]

Epilepsy and deviance were similarly entwined at the White Special School. Beginning in the early 1940s, administrators routinely admitted for electroencephalographic observation applicants who had "never given any clinical evidence of seizures, but [who] may have periodic outburst of temper."[132] Teachers at the school described the "bizarre behavior" of these primarily male students, seemingly inclined to episodes of "minor or major hysteria," in contrast to the "emotional control" exhibited by "easier to deal with" students with grand mal seizures.[133] In 1943, the school opened its Psychiatric Room exclusively for such "problem boys" and instituted regular electroencephalograms in an attempt to understand whether epilepsy caused their troubling behavior.[134] Results by the mid- to late 1940s suggested that most students in this group shared "a form of psychomotor epilepsy" otherwise characterized by unusual sensations and involuntary body movements. Their periods of altered awareness and sometimes automatic or unconscious behavior, also cited in cases of "epileptic criminality," were therefore deemed "dysrhythmic" in nature and responsible for their frequent misconduct.

In his study of brain waves and his writings about epilepsy, Dr. Lennox also searched for electroencephalographic evidence of the seeming connections between epilepsy, social deviance, and criminal behavior. As early as 1941, one of his interests in the White Special School, in fact, had been to "take a random sample" of its "severe problem boys" for comparative research.[135] As he explained to the school's Superintendent Metzner, among the World War II draftees he studied, those with "a history of jail sentence, family history of seizures, a personal history of head injury, etc." had "abnormal records as high as 40 to 80 per cent."[136] He anticipated similar findings in a "sample of the prison population" he would soon examine.[137] In contrast, "normal family histories" among those draftees Lennox deemed to have "abnormal" EEGs were nominal.[138] According to his preliminary findings, issues of criminal behavior and cerebral dysrhythmia appeared to be undisputedly, if not bioelectrically and genetically, linked.[139]

Furthermore, despite EEG's established place within neurology by the late interwar period, its use by postwar psychiatrists and psychologists continued to connect epilepsy to studies and pathologies of the mind. Since Hans Berger's earliest tests in the 1920s of patients diagnosed with schizophrenia and hysteria, EEG had been employed by a small group of researchers to understand the electrophysiological features of human emotions and mental states. Concurrent with Lennox and the Gibbses' electroencephalographic work with epilepsy in the 1930s, for example, was that of their Harvard colleague Hallowell Davis, who used EEG to examine brain-wave changes in patients undergoing psychoanalysis.[140] Lennox's own findings by the early 1940s also suggested a bioelectrical resemblance between "mental disease" and the propensity for seizures. Tracing an intrinsic relationship between people with epilepsy and those with psychological disturbances, he reported that many schizophrenic patients and problem children alike revealed "transient dysrhythmias."[141] New understandings in military aviation that "washouts" possessed "epileptoid waves" similarly linked epilepsy to delusions, panic, and an unstable psyche, while criminal trials featuring EEG suggested that seizure-like bouts of "temporary insanity" might produce episodes of extreme violence.[142]

Epilepsy's associations with sex and sexuality also expanded after World War II with electroencephalography's integration into postwar sex research. The technology was featured for the first time in American biologist and sexologist Alfred Kinsey's *Sexual Behavior in the Human Female* in 1953, the highly anticipated sequel to the 1948 bestseller *Sexual Behavior in the Human Male*.[143] As a new tool for understanding the neurophysiology of sex, electroencephalographic research in the 1953 volume confirmed a "most remarkable parallel" between epilepsy and sexual response.[144] Namely, the dysrhythmias found in epilepsy provided a baseline of scientific understanding for "normal" and "average" brain activity during sex. Seizures became scientifically analogous to orgasm, and female orgasm in particular.[145] Kinsey based these conclusions on research by Argentine psychiatrists Abraham Mosovich and Alberto Tallaferro, whose broader project on psychoanalysis and epilepsy had led them to make EEG recordings in masturbating subjects.[146] According to their results in "normal" male and female participants, the pair was "tempted to establish a similarity with what happens during an epileptic convulsion."[147] Records during sexual climax in particular revealed that "the phase of mechanical tension corresponds to the tonic phase of a seizure which gives way, during unconsciousness and clonic convulsions, to the relaxation and release of tension."[148]

When Kinsey first contacted Mosovich in 1952, the two shared the belief that the entire nervous system, including the cortical and subcortical structures, was involved in all stages of sexual response.[149] EEG, in Kinsey's opinion, was therefore an untapped "measurement of very considerable scientific importance," and epilepsy, an "essential" unit of comparison.[150] In attempting to consider the participation of the whole nervous system in sexual climax, however, and "not merely," as Kinsey explained, "the limited genital area which has been at the center of most previous studies," the method also reflected perceptions that orgasm was more generally "not unlike epilepsy."[151] Although only in "rare cases" did seizures involve the musculature of the genitals, Kinsey stated in the *Human Female* that the "resemblances between the spasms of convulsions in epilepsy and those following orgasm" were undeniable.[152] Noting the "obviously involuntary" movements and "convulsions" of each, he further cited observed comparisons dating as far back as Democritus in 420 BCE.[153] Like French neurologist Jean-Martin Charcot's more recent sexual underpinnings for "hystero-epilepsy" among nineteenth-century women or Sigmund Freud's subsequent theorization that seizures were "exaggerated tensions of the libido," Kinsey read existing analogies present in the culture through a new technological lens.[154]

Moreover, despite Mosovich and Tallaferro's examination of both male and female subjects, or the fact that their research happened to correspond with the timing of the second rather than the first of Kinsey's volumes, it was "female orgasm" that EEG and epilepsy seemed to particularly illuminate. With its purportedly "vague" physiological markers, female sexual response had a long history of being dismissed by researchers owing to perceptions that it either did not exist or remained inessential to the reproductive "purpose" of sex. Electroencephalography's apparent ability to read the mysteries of women's bodies directly through their brains was thus appealing when seeking insight into female subjects. Reflecting on the value of their research, Kinsey congratulated himself and Mosovich for battling institutional prudishness, stating in a 1953 letter, "Our tradition against such observations [has] . . . done us a great deal of harm and I think the more we learn on this subject the more benefit man will get from it."[155] By associating female sexual response with seizures and vice versa, however, the pair's work arguably pathologized women's sexuality further while also sexualizing epilepsy and its historic "tensions of the libido" in new ways.

Other postwar electroencephalographic studies served to align epilepsy with scientific and mystical notions of genius. During World War II, advocates intent on elevating epilepsy's profile had cast people of historical significance, including Socrates, Joan of Arc, Napoleon, and Vincent van Gogh, as "epileptic geniuses."[156] In popular coverage after the war, including the 1953 book, *Genius and Epilepsy*, explanations of the confluence between epilepsy and genius came to lie in brain waves: "A study of epileptics who were great reveals that the epileptic is in some ways in a favored position for mental development. . . . What can account for greatness among some epileptics? For one thing, the variations of the *brain-waves* of the epileptic are wider than those of a normal person."[157] It continued:

> If it is his lot to have the "falling sickness," it may be his privilege, also, at times, to have the rising inspiration. If he must go deeper into the valley of despair than does the "normal" person, he may also rise to unusual peaks of hope, and vision and courage. If consciousness must at times be partially or completely obliterated, it may at other times reach a peak where it beholds facts, intuitions, relationships, not given to the normal person to experience.[158]

Such interpretations echoed older explanations of epilepsy's crashing "highs and lows" and the transcendence of seizures. New, however, was the suggestion that the "valleys" of adversity paired with the "peaks" of prodigious insight that ostensibly formed or fostered genius mapped directly onto the contours of brain waves.

Indeed, brain activity during seizures and epileptic auras was long believed to offer some people with epilepsy an unparalleled, generative state of creativity if not a connection to the world beyond the ordinary. This was particularly true in cases of lost consciousness and the convulsions seen in grand mal seizures. EEG and other twentieth-century studies of the brain provided new modes of investigation for these suppositions. Interwar researchers engaged in electroencephalographic sleep studies at Tuxedo Park, for instance, were equally interested in what brain waves revealed of dreams and the psyche.[159] Wilder Penfield's electrical stimulation of epilepsy patients' temporal lobes, which he understood to imitate the electrical impulses of seizures, had at times produced hallucinations that bordered on visionary (see chapter 5).[160] The alleged "other side of the seizure," which was to say its connotations beyond impairment or decline, resonated with characterizations of divine inspiration found in countless depictions of the "falling sickness" since antiquity. The application of EEG, however, helped write new and earthly explanations for these understandings. Unlike the bioelectrical "monotony" that explained most people's flattening "normality," geniuses of all kinds produced highly "active" brain waves like those observed in epilepsy.[161] As the *New York Times* reported of eminent mathematicians Albert Einstein, John von Neumann, and Norbert Weiner in 1953, "very active brains, such as Dr. Einstein's while he was thinking of his theory of relativity," achieved such patterns "very rapidly."[162] In "brainstorms" and strokes of insight, epilepsy and genius thus mirrored and perhaps even informed one another.

The Waning Eye

Electroencephalography's presence in science and medicine only increased throughout the postwar era—a period, historians show, that coincided with growing enthusiasm for new technologies and expanding medical research. Nevertheless, ambivalence regarding EEG's powers of perception also began to show in this period.[163] Early in its history, the expense of EEG machines and their need for expert operation, interpretation, and care had put the device and technique beyond the reach of even most hospitals. Neurologists like William Lennox accordingly acknowledged that the value of EEG was in many cases "more scientific than practical."[164] Yet whereas practical limitations in the 1930s and 1940s were the primary reason that EEG might be overlooked, by the 1950s, many early proponents, including Lennox, questioned the extent of what it actually captured.[165]

Staff at Detroit's White Special School, for instance, seriously considered retiring their once celebrated EEG machine in this era.[166] While the updated model they received in 1948 had addressed administrators' initial concerns about frequent breakdowns and ink spills and the "unsightly" copper grounding cage that "on many occasions frightened children," a more fundamental challenge to the device's seeing capabilities arose.[167] A 1953 report to the school board showed growing skepticism: "We feel that an EEG is a very small part of any diagnosis and taken alone has little or no meaning."[168] "Diagnosis," the report continued, "used to be formed on the basis of whether the reading was 1, 2, 3, 4, or 5. We feel this type of diagnosis is not valid. . . . EEGs are not always perfect and without further neurological study no true picture can be formed."[169] Again in 1955, only one year before the school closed as an institution for students with epilepsy alone, administrators expressed measured praise for the technique, stating, "Like other things, it is not infallible, but it usually shows if something is wrong."[170] By the middle of the decade then, much of the original conviction among such early adopters had been replaced with a degree of uncertainty and a sense that EEG's capabilities were limited in scope.

Concerns about the lack of a "true picture" in Detroit echoed broader opinions by the 1950s that EEG inhibited more holistic forms of diagnosis and understandings of disease. Several physicians and people with epilepsy argued that an electroencephalogram provided a partial view and that epilepsy was a condition involving much more than a person's brain waves. Having rendered a somewhat reductive picture—making brains, for example, as comprehensible as "household gadgets" and both "bad" and "sick" kids simply "dysrhythmic"—neurologists emphasized the importance of other types of evaluation. Dr. Lennox, though his efforts helped place EEG machines in hospitals and neurological units across the country, became more tempered in his endorsement. Popular depictions likewise set aside some of their initial hyperbole. Whereas in 1937 EEG had been the unmistakable "test of identity" and "index of the brain," comparable text by the mid-1950s conveyed greater nuance: "The electroencephalograph in no sense indicates how we think, but it does show how sensitive the brain is."[171] The seemingly boundless potential of the technology to see, including the ability to perceive something definitive of human identity, all but disappeared.

Contributing to this change, perhaps, was the fact that electroencephalography failed to address the ambiguities associated with diagnosing an intermittent and variable neurological condition like epilepsy in the first place. Despite its apparent capacity to test for the dysrhythmic patterns indicative of epilepsy, EEG in some ways produced more of the uncertainty that doctors and researchers intended it to mitigate. For one, brains and brain waves were never as self-evident as once claimed. Electroencephalograms required a trained reader, and at outposts like the White Special School, where technicians but few neurologists were available to read results, not everyone possessed or could achieve literacy of the brain.[172] Moreover, it was not always clear what constituted epilepsy. Phenomena such as "borderline" cases and "questionable 3s" remained, as did the interpretive nature of the results. Electroencephalograms also rarely revealed the unknown but "clearly epileptic" cases of epilepsy that had captured research and public interest early on. Equally, for those who did not appear to have epilepsy but were implicated by a brain-wave test, the diagnosis of "cerebral dysrhythmia," "carrier," or "subclinical epileptic" proved a confusing and rarely adopted expansion of the diagnostic label when seizures did not follow.

For these reasons, as well the material and technical constraints that proponents identified, EEG was scarcely used to "screen" for epilepsy in the ways once promoted. Of the underused premarital electroencephalogram, Lennox reasoned that "very few medical centers are yet equipped with the proper apparatus, or . . . doctors skilled in the interpretation of the records."[173] The practice, however, was also likely impeded by a lack of motivation among individuals to undergo costly and potentially life-restricting testing to begin with. Similarly, EEG was never implemented as a stand-alone test of driver suitability. Even as people with epilepsy were allowed to drive under certain circumstances in some states after World War II (see chapter 4), physicians generally wanted to adjudicate on a case-to-case basis, fearing that an electroencephalogram would occasionally approve the wrong candidate while eliminating others who were safe to drive.[174] Correspondingly, EEG failed to become a definitive measure of epilepsy or criminal intent in various legal trials across the country. Likewise, as historians note, use of the technique in the US Air Force was short-lived, with EEG rejected as "not sufficiently promising" after 1945. Given the poorly articulated correlation between "epileptoid waves" and emotional control, as well as the technology's narrow focus on the individual's physiology over his environment, EEG was soon cast aside.[175]

Ultimately, the identifying impulses expressed through electroencephalography in the 1930s and 1940s were replaced in the postwar era by a more equivocal stance—one that, in addition to acknowledging EEG's limits, reflected a rising vision of people with epilepsy as "near normal." In the mid-1930s, epilepsy seemingly became comprehensible through brain waves, making proponents like Lennox the "good physicians in modern dress" who "beat the devil out of epilepsy."[176] By the mid- and late 1950s, however, rather than providing certain and clarifying evidence, brain waves often complicated, or indeed oversimplified, the picture according to some observers. Though epilepsy was discussed with rising frequency in public forums, its general lack of disclosure beyond clinical and private contexts made the prospect of exposing those with epilepsy, or mere "carriers" of dysrhythmias, less desirable. Thus, from its early applications in epileptology to its expanded uses and contestations, EEG's powers of sight changed too; within a relatively short period of time, the technology emerged and retreated as one of superlative sight and detection. While EEG would forever change what it meant to diagnose and have epilepsy, its ways of seeing also exposed the growing difficulties of revealing a "secret illness" while simultaneously establishing new expressions of that very idea.

II   MANAGING RISK AND SAFETY

3

Safe Citizens at Detroit's School
for Children with Epilepsy

In the spring of 1946, administrators at the White Special School—the nation's only public school program for children with epilepsy—removed the rubber padding from its many radiators and stairwells. Reinterpreting the previous decade's measure as counterproductive to student health, educators at the Detroit-area school had abruptly reversed several of the protocols once relied upon to insulate children and their seizures. For the first time in the institution's short history, students moved freely through the hallways, carried their lunch trays in the cafeteria, and even wandered in public settings on newly arranged fieldtrips. Unassisted by safety apparatuses or helping hands, postwar pupils moved to engage the commonplace perils of their physical and social environments, approaching situations and places once avoided for fear that a seizure might occur. In a refrain made popular among teachers, board members, and the school's medical personnel after 1946, gone were the days of the "protected" child with epilepsy.[1]

The sudden outmoding of safety protocols at the White Special School manifested changing values about epilepsy and its management in the decades after World War II. When the school first opened in 1935, its primary aim had been to provide those children with epilepsy excluded from regular schools—regardless of race, class, ethnicity, or religious background—with both educational instruction and medical care. Within the school, an array of structures and practices shielded students from possible falls and other harms, facilitating a particular vision of child safety and the institutional containment of a disability. Yet soon after the war, the belief that new anticonvulsant drug therapies had brought seizures under control compelled school leaders to purge these measures and declare its students "safe." Seizures, however, persisted; even as the rubber was ceremoniously stripped from the school's fixtures and the outside world was deemed an extension of its playground, there was little reduction in the seizures that occurred on the school's premises each day.[2] Consequently, when the school abandoned its interwar program of protection, it did so as seizures remained to be managed and as a diagnosis of epilepsy continued to present challenges in American society.

This chapter examines a little-known social experiment involving seizure-prone schoolchildren in mid-twentieth-century Hamtramck, Detroit. It explores how a diverse group of participants at the White Special School (the "White School" for short) attempted to redraw the physical and definitional boundaries of safety to rehabilitate and socially integrate children with epilepsy. In practice, this postwar transition involved eliminating the external measures once trusted to protect students from seizure-related injuries and replacing them with a broad set of internalized safeguards. By the time the school closed in 1956, the assumption was that children would not only move to "regular" schools but do so as minimally disabled, and perhaps even invisibly disabled, individuals whose epilepsy was privately managed. In this way, the school's treatment of epilepsy—and indeed, the very locale of safety— shifted from the institution and its hardware to the bodies of the students themselves.

This chapter calls attention to how children's safety and children's environments operated as sites and strategies for expanding social citizenship in this period.[3] Within histories of childhood and education, school settings feature prominently in analyses of how young citizens of receptive age have been molded into ideal models.[4] For children with disabilities, historians further note that growing special education programs throughout the twentieth century tended to consign disabled children to more limited geographies and forms of belonging—particularly before, but also after, efforts to integrate them into mainstream education were successful in the 1970s.[5] The unexamined history of the students with epilepsy at the White Special School suggests a different trajectory; owing in part to growing characterizations of epilepsy as an occasional and therefore minor, disguisable disability after World War II, the school aimed for complete assimilation. As seizure-prone children became seen as resilient, contained, and, not least, impervious to injury, they also became "normal" and thus eligible for a degree of honorary inclusion denied to those whose more visible differences kept them separate for decades longer.[6]

The case of the White School also challenges dominant narratives about the ways that epilepsy and people with epilepsy went public after World War II.[7] Although administrators and school physicians emphasized Dilantin and other medical interventions as the means to increase children's involvement in community, work, and education, the changes that unfolded at the school were only partly medical in nature. The postwar period was not the first time that children or adults with epilepsy had participated in such arenas, nor were seizures or their social risks eliminated by new drugs. Such medical developments did, however, increase expectations that people with epilepsy would live and present as mostly seizure free.[8] By contrast, the work of reimagining seizure-prone children as safe and self-contained in this period, as first implemented at this public school, shows how these goals were forged upon and through the ostensibly malleable bodies of children and the spaces they inhabited in the decade after the war.[9]

Changing definitions of safety at the White School further complicate understandings of the immediate post–World War II era as one of growing insularity and protection around "disabled" and "normal" children alike.[10] At the very moment that protective measures for most children were tangibly and rhetorically expanding—including on playgrounds and in classrooms— the mechanisms that buffered those with epilepsy were intentionally removed.[11] For children responsive to medical treatment, the visible markers of their disability quickly became a liability exceeding most physical seizure risks. In the safety-saturated but seizure-adverse context of postwar America, concerns about the potential dangers of epilepsy were thus subsumed to larger efforts that sought to rebrand children with epilepsy as safe citizens. Such safety was defined not by protectionism but by the promise of a child's "normalcy" and their ability to function as essentially "non-epileptic." This included, among other things, an adherence to dominant gender roles, future participation in the industrial economy, and the visible expression of independence as achieved through medical self-management.[12]

Finally, this case underscores the extent to which the outcomes of these efforts were unevenly shared. In addition to students whose multiple and more noticeable disabilities assigned them to special education programs well past the school's closure, African American children schooled and rehabilitated alongside white peers at the school moved to an effectively segregated system.[13] Girls, too, like their peers in regular schools, rose to different standards and were not directed toward the physicality or career orientation expected of male students. Eliminating the stigmatizing signs of one's disability, in other words, did not afford Black children the same privileges as their white counterparts or give girls the same opportunities as boys, nor did it free any child from the constraints of neighborhood or social class. Though an educated and rehabilitated child with epilepsy may have been able to minimize or even invisibilize his or her epilepsy by controlling seizures, many would not reach the level of inclusion prioritized for white, "mentally normal," and otherwise able-bodied children. All children at the school, however, faced new pressures to achieve self-control and integration—qualities that placed an onus on every seizure-prone child to strive for more while simultaneously requiring less from those around them.

Origins of a Protective Experiment

The White Special School was founded in 1935 by the Detroit Board of Education as part of a broader Social Security Act initiative to expand educational programming for children with disabilities. Detroit at the time had a long history of providing special education to its "deaf, blind, and crippled children," yet those with epilepsy were notably absent from such efforts.[14] In addition to institutionalization and the numerous exclusions that people with epilepsy faced throughout the late nineteenth and early twentieth centuries, the presumed mental and emotional effects of epilepsy had barred most from classrooms across the country.[15] Described as egocentric, explosively irritable, and mentally dull, children with epilepsy were routinely dismissed from schools or kept home by parents. Because epilepsy was associated with mental deficiency, many children were considered unsuitable pupils and classmates.[16] Seizures, moreover, were believed to present challenges for the peers of epileptic children. As a national school health report concluded in 1928, "no child who has attacks of epilepsy should be permitted to attend school," as the "effect upon other children is too demoralizing."[17] School seizures were thus grounds for expulsion and could inspire recommendations for institutional placement. Prior to the opening of the White School, most children with epilepsy, if educated at all, received their instruction at home.[18]

For members of Detroit's Board of Education, however, committed to

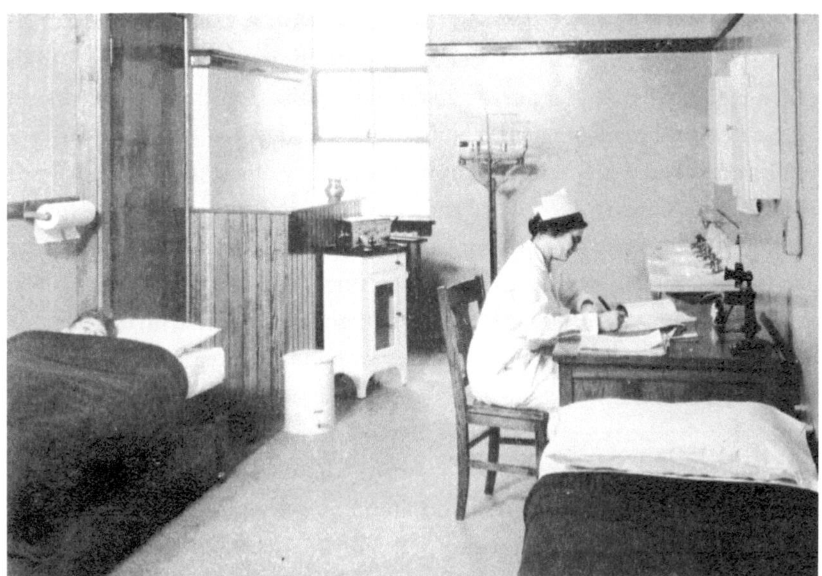

Original caption: "Continuous nursing is a part of the twenty-four-hour service in the school for epileptics." Superintendent's annual report for Detroit Public Schools, 1937. Walter P. Reuther Library, Archives of Labor and Urban Affairs, Wayne State University

special education and resolute that "every child should be trained within the limit of his capacity," the White School aimed to bring underserved seizure-prone children into the public system.[19] Given Detroit's history of special education and resources, they also regarded their city as the ideal location for such a program. In 1936, the White Special School established headquarters in a wing of Katherine B. White Elementary, a primary school in the predominantly Polish enclave of Hamtramck in the northern part of the city. Detroit was then a prospering center of industrial manufacturing that was undergoing massive expansion. It was also home to an increasingly racially and ethnically diverse population.[20] Variously disabled children from neighborhoods all over the city were connected to its special schools by a robust network of mobile coaches. Initially functioning on a residential model, the White School for a short time operated as a five-day-a-week boarding school before converting to a day school in 1939.[21]

Those admitted to the school in the 1930s were by all accounts "highly selected" and carefully calibrated to the institution and its aims.[22] Initial appointments were made through the school board at the request of principals, agency workers, and parents. Its on-site clinic, comprising a neurologist, a psychiatrist, and a nurse, would then evaluate each candidate based on medical and psychological examinations and detailed family histories.[23] While an accepted candidate might have a history of grand or petit mal seizures or the "unusual behavioral patterns" seen in psychomotor epilepsy, a diagnosis of epilepsy was not enough to secure admission. Applicants who had IQs of less than 65, who exhibited nocturnal or nighttime seizures only, or whose seizures were deemed "too infrequent" were rejected on the grounds that they were "ineducable" or unlikely to require the care offered by the school.[24] Those admitted, however, were seen as a representative sample of the city. Students were mostly white and American born, though several were the children of European immigrants; there were roughly equal numbers male and female students, and approximately one-fifth to one-third were African American. The school's racially integrated program, though a highlighted feature of the program, was not unusual for institutions for sick and disabled children at the time.[25] By the end of World War II, the school's population rose from its original class of 48 students to 194 students, a figure that remained relatively consistent until the school's closure in 1956.[26]

During its early years of operation, the educational goals of the White School were made secondary to those of fostering student health. Children received basic instruction in reading, writing, and arithmetic, yet a regimen of "hygienic living," including a simple diet and outdoor exercise, formed the basis of daily instruction and care.[27] The school also became an early testing ground for a number of experimental treatments co-sponsored by Detroit's Board of Education and Department of Health.[28] The first class of 48 students, for instance, were part of a two-year study with the Detroit-based pharmaceutical company Parke, Davis to test the seizure-reducing effects of sheep brain lipoids injected intramuscularly. The 1935 study, monitored by staff neurologist O. P. Kimball, reported mixed but favorable results, with 12 percent of students "cured," 50 percent "improved," and 37 percent exhibiting "no improvement."[29] Moreover, once it was deemed "O.K. for oral use," but a full year before its approval, White School students were among the earliest recipients of what eventually became the first mass-marketed anticonvulsant drug, Dilantin.[30] The medication, which came to replace many of the barbiturates and special diets used to control seizures before 1938, was the first of many supplied and tested at the school and, indeed, required for attendance. While Dilantin and others went on to wide acclaim, some medications were later discontinued as a result of dangerous side effects, worsening symptoms, or inefficacy.[31] Except for a few instances of noncompliance, most children and parents were said to be "enthusiastic" about the treatments.[32] Though potential cures and new methods of seizure control were not the express purpose of the school, interest in such developments promoted a willingness to experiment on children, making what might otherwise be deemed a risky practice integral to the program. Likewise, the school was among the earliest sites in the country to use EEG in admission screenings, routine checkups, and drug testing. Though the procedure could be intimidating for children, it quickly became a central component of the school's clinic.[33]

Beyond diagnostic and therapeutic services, however, the White Special School worked primarily to manage the many seizures that occurred on its premises each day. The protective protocol around grand mal or convulsive seizures was particularly well developed, with devices and trained staff positioned to mitigate their impact at every juncture. Coaches, all motorized by the 1920s, delivered pupils between home and school. Attendants also kept "continuous watch" over children in the schoolyard, while teachers with graduate-level training in epilepsy and its management, mostly women, taught classes capped at twenty-five students.[34] Supportive technologies to "guard the children's safety" included high-back canvas chairs and wraparound desks in the classrooms, "heavy, resilient" rubberized coating on hallway stairwells, radiators, and railings, and ample padding in the interior of the coaches should a student have a seizure during the commute to or from school.[35] A buzzer system connected the school's rooms and corridors with the infirmary so that a nurse could promptly assist when a seizure began. Children were both protected and protectors in this network; a "buddy system," which required students to travel between classes and to the lavatories in pairs, was established to reduce unattended seizures. When seizures did occur, pupils went to the nurse's office, where they rested on cots under full supervision for as long as it took for them to recover.[36]

The information that doctors, administrators, and nurses collected about students in this early period also contributed to a model of vigilant protection. Nurse Christine Deeward, who kept detailed medical records for each student, was responsible for recording all seizures. She also compiled this data into broader statistics and graphs deemed useful for school policy.[37] In

Early classroom at the White Special School, 1937. Note the high-back chairs. Walter P. Reuther Library, Archives of Labor and Urban Affairs, Wayne State University

addition to revealing possible trends in individual students' seizure patterns, tracking of this kind was said to have predictive and thus preventative value. Seizure data coming out of the school's clinic, for instance, quickly reinforced existing wisdom that children with epilepsy should not be "overtaxed" physically or mentally. When, for example, the highest numbers of seizures were found to occur in September and May of each year, staff concluded that the physical and psychological pressures of beginning and ending the school year induced seizures.[38] As a result, until after World War II, students carried a half-semester's workload throughout the year so as to reduce the overall number of seizures seen at the school.[39]

For reasons of both health and safety then, the expectations placed upon students—educational or otherwise—remained modest throughout the 1930s and into the 1940s. As evident in the school's mission statement in 1941, it focused more on stabilizing students than on promoting their scholastic achievement or wider social development. "We have endeavored to train this group of handicapped children to meet their deprivation with dignity," it explained, "to have a perspective in seeing themselves in relation to the universe, and to, in the knowledge of their affliction, extract the maximum pleasure from the world in which they live."[40] In 1941, the environment in which such children lived and operated was still assumed to be small, and appropriately so. Although parents, teachers, and administrators discussed the possibilities of future work and living arrangements for children, their "world" was also tightly regulated. The single largest consideration was to minimize the impact of seizures and thus to keep students both physically and socially insulated in a safe place beyond their homes.

The Postwar Dissolution of Protectionism

Immediately after World War II, both the White Special School and epilepsy became subjects of unprecedented publicity.[41] Around the time the school was establishing itself, the first epilepsy advocacy groups had begun to organize, declaring epilepsy both a deeply misunderstood and a medically neglected condition.[42] The overall visibility of disability also expanded during the war, following the increased participation of adults with disabilities in the domestic workforce and a rising government focus on disabled veterans.[43] As soldiers returned from war with epilepsies acquired in battle, neurologists and public health officials voiced further concern about the lack of medical and occupational opportunities available to such men.[44] Children were a parallel source of anxiety; amid a growing emphasis on child health and welfare, the recognition that children were disproportionately afflicted with epilepsy and that most conditions emerged by age six dominated discussions of what was quickly becoming a public condition.[45]

Particularly powerful in messaging about epilepsy after the war, for both children and adults, was the role new medications would play. A number of magazine articles and health pamphlets heralded anticonvulsant treatments as the means to public acceptance, while the White School echoed that epilepsy was now a "hopeful condition" owing to the very drugs that continued to be tested on its premises each day.[46] Featured in *Life*, *National Parent-Teacher*, and *Coronet* magazines in 1946, the school soon became an exemplar of progressive scientific treatment, drawing visitors and written inquiries from near and abroad. Parents wished to relocate to Detroit and enroll their children at the school, educators looked to initiate similar programs in their own cities, and people with epilepsy asked about the latest medications and employment opportunities.[47] In a flurry of activity and with a public image of its own to suddenly manage, the White School began to redefine itself and its pupils from the inside out.[48]

It was in this context of increased hope and visibility, in fact, that the school rejected its previous methods, deeming its earlier program ineffective and unmodern. Neither enthusiasm about new drugs nor the school's rising profile, however, meant that seizures were any less prevalent at the school. Decisions to remove various safety features in 1946, such as rubber stairs, required instead a new system of safety. Administrators drew upon several less overt tools to justify and indeed create these changes. They mobilized new social scientific discourses regarding child "overprotection" and self-determination to fortify the image of students. They also rebranded seizures as "safe" and seizure-prone bodies as essentially resilient. Finally, they moved children with epilepsy into public spaces and toward new types of social participation. Since much of this change was predicated on the promise of medicinal seizure control, it in many ways materialized a shift that was more aspirational than real.

The White School's rising critique of protection and "overprotection" had its basis in interwar child psychology, which gained increasing momentum at the school during World War II.[49] Having consulted with child psychiatrists and psychologists from its founding, the school had been responsive to the growing authority of psychoanalysis in nearly all matters of child-rearing and early education.[50] Unlike concurrent theories regarding "refrigerator mothers," which placed a lack of parental warmth at the root of childhood autism and schizophrenia, theories about overprotection posited that parents and caregivers could, and often did, overwhelm children with anxious care. Reflecting broader debates about "smother love" for disabled and "normal" children alike, White School psychologists by 1943 noted that "too many boys and girls are coming from homes as pampered over-protected individuals."[51] Unhealthy family dynamics, as well as inappropriate attachments to parental figures—most commonly "nervous" and "overprotective" mothers—were cited as both the source and perpetuating cause of epileptic disorders and poor social adjustment.[52] Further associated with juvenile delinquency, epilepsy was equally understood to be an illness of broken and dysfunctional homes in which children were shown inadequate love or care.[53] In both cases, experts described family fear and ignorance as the "real handicap," which prevented "satisfactory adult citizenship" later in life.[54]

From the beginning, such theories encompassed the physical body and psychological state of the child with epilepsy, regarding the two as inextricably linked. As school administrators explained by 1946, "Many of our beliefs concerning the care and welfare of the epileptic child are undergoing a transition. We know that these children are less likely to have seizures when leading a happy and active life, with only enough protection for their own immediate safety, and that limiting activities plus over-protection is likely

Students in front of the White Special School, including a girl helped by a medical attendant following a seizure. Detroit, MI, ca. 1946. Photo by Mark Kauffman. Mark Kauffman/The LIFE Picture Collection/Shutterstock

to produce neurotics, dull self-sufficiency, and warp independence."[55] According to the report—the same that celebrated the removal of safety features at the school—measures aimed at safeguarding students' bodies and minds were not only unnecessary but undeniably harmful. Exposure to everyday "dangers," by such logic, also stood to reduce seizures. An ethos of equal subjection to life's hardship and challenges thus emerged as administrators, teachers, and medical staff came to encourage modest degrees of physical and emotional risk for students. "Coddling" and "pampering" were to be strictly avoided, not only because they might limit a child but because needless protection would reinforce, and in some cases encourage, seizures.[56]

Despite the growing blame on mothers and families, parents by all accounts supported the school's efforts to reduce what it called protectionism both at the school and in the children's homes and family lives. In 1947, the school launched its Parents Group—proudly noted as "the first and only organization of parents of epileptic children in the country"—to strengthen communication between families and staff with regard to these and other matters.[57] Through meetings and committee work, participants focused on reducing epilepsy's stigma, raising awareness, and creating opportunities that would increase student independence. Describing the "therapeutic value" of the group, faculty reported that several parents felt relief in interacting with other parents of children with epilepsy, and that "talking" rather than "hiding" the condition reduced parental feelings of shame and isolation.[58] Equally noted was how a shared concern for seizure-prone children created a sense of community, fostering cooperation across lines of race, class, and neighborhood represented at the school. Parents of Black and white students, formal reports stated, worked together at integrated picnics and fundraisers and came to speak of "our boys and girls" rather than "my child."[59]

Equal to these efforts were attempts to bolster the self-esteem and self-awareness of students with epilepsy. In 1945, the school implemented individual counseling and a new class called "Personal Relationships," in which the children were invited to "discuss their personal feelings in an accepting atmosphere."[60] In class, students reflected on their individual experiences and struggles as a group with some shared identity.[61] As characterized by teachers and administrators, students communicated primarily a "common feeling of being deprived by their families of social opportunities."[62] They also expressed feelings of rejection as "the only afflicted member of the family" and "a general air of protest against what they considered overprotection."[63] According to reports, the free exchange of such feelings transformed dynamics not only in the classroom but within families as well. As students gained confidence, they also created their own organizations, including a group for older children called the "EEGs," standing for "Epileptics' Educational Group," which worked to change public perceptions that people with epilepsy needed "special treatment."[64]

The idea that students required no such special treatment, and that they could and should in fact exercise a high degree of independence and self-sufficiency, fit with growing discussions of social citizenship at the school. Evoking what Dr. William Lennox, neurologist and friend of the White School, identified as "civilian epileptics" during the war, postwar administrators sought to emphasize the intrinsic value of students and their capacity to lead "full" and "productive" lives.[65] Lennox's formulation had referred to the "regular" and "respectable" men and women who suffered quietly with epilepsy—those who were "physically and mentally normal" and "eager to work" but who faced substantial prejudice and rejection when their condition was discovered.[66] Administrators' emerging notion of students as young citizens similarly encompassed a desire for social acceptance and opportunity while simultaneously signaling something more ambitious. Namely, talk of citizenship implied that those children closest to "regular" and "respectable" could be groomed to achieve a level of "normalcy" that was previously unheard of for seizure-prone people, thereby redefining what it meant to have epilepsy for a new generation of Americans.

To forge a path to greater social belonging, White School administrators emphasized the importance of a child's individual development, expression, and separation from the family.[67] While the psycho-emotional pressures of overprotection were deemed damaging, the "regular pressures" of daily life were reassigned as positive and even necessary. A new curriculum promoting student exploration after 1945 became vital to the program.[68] In art and drama classes, children were encouraged to channel their negative emotions through creative expression. Younger students listened to music and engaged in therapeutic drawing and dance. "Rhythm records," for their "strong and instinctive beat," were considered especially beneficial for fostering movement and the free release of emotion.[69] Once subsumed to medical considerations, academic achievement also became a point of focus. The school, for instance, abandoned its "never under tension" rule and the idea that mental exertion caused seizures.[70] Rather than being treated as "slower and sicker," with a curriculum tailored to half-pace, students were held to the same standards as their peers in the regular school system. By the postwar era, the school accommodated all grade levels and encouraged high school graduation.[71] Citing the importance of equal expectations, administrators explained, "We do not believe in eliminating competition. . . . The same pressures should be placed on an epileptic child as on any other child"—be it at school or in their future career.[72]

In addition to increasing intellectual demands, postwar students were expected to participate in a growing number of extracurricular activities and life-planning exercises. Student council, spelling bees, and various clubs invited children to develop their social skills and practice adult roles. Mirroring the rise of vocational rehabilitation programs after the war, a new course called "Career Development" encouraged students to consider their employment prospects. The involvement of men with disabilities in Detroit's wartime workforce, as well as open hiring practices for people with epilepsy at the Ford Motor Company in the 1930s, provided local precedents for imagining such work.[73] Parents' meetings similarly focused on issues of vocation. Fathers, employed in various industries across the city, discussed job opportunities for their sons and possible office or hospital work for their daughters. Some also found work placements for one another's children as teenagers.[74] Whereas the prewar curriculum had emphasized basic learning and light work therapy such as gardening, students, and especially male students, were now encouraged to imagine a lifelong occupation.[75]

Increasingly, administrators also argued that seizures and the accidents that sometimes accompanied them should not be feared by students, parents, or staff at the school. The buddy system was consequently among the earliest of prewar measures to be eliminated after World War II. As Principal Harper Hazel explained, "We decided—what if a child did have a seizure in the hall? He couldn't hurt himself and someone would see him and ring for the nurse or attendant." Following the changes at the school, she noted, "children went between classrooms, to the office, on errands, or to the lavatory. So far we haven't regretted the changes and the children are much happier."[76] Along with the buddy system, the school retired the rubber coating in its hallways and classrooms and allowed children free movement in the cafeteria, including an insistence that they uncap their own milk bottles without assistance.[77] Students who had seizures after this time, moreover, were said to "recover quickly," needing no more than a short nap in the nurse's office rather than a full day of rest. Thus, while children's physical safety and broader protection from nuclear war and other threats in this era became rising points of concern, those who experienced seizures saw some of the safety measures surrounding them lessen. Many, in fact, saw expansions of their personal freedom.[78]

In 1947, school administrators compiled a key report titled *Are Seizures a Safety Hazard?* that further supported these changes.[79] Prompted by "the many questions directed to us concerning the accident ratio of epileptics," they sought to address "misconceptions" about the danger of epilepsy on the job and in public.[80] Serving as a laboratory for wider debates about employment practices after the war, the school collected and compared data on all accidents resulting from seizures with those that happened as a matter of course. Of the 2,600 grand mal convulsions that occurred at the school over the preceding five years, the report found that only seven had produced accident or injury "serious enough to report."[81] These included two incidents of "broken front tooth," and one each of "wrenched back, bruised knee and ankle, and broken clinical thermometer in mouth."[82] More dire and numerous than these few incidents, however, were the injuries resulting from types of accidents that "could happen to anyone."[83] The bodily harm incurred in thirty-seven instances of nonconvulsive falls, art room mishaps, imaginative play gone awry, and bites from wayward dogs, for example, had demanded greater medical attention than those incurred during seizures. With "regular child's play" proving more dangerous than epilepsy itself, the report—subsequently distributed to "interested organizations" of government and industry across the country—provided a larger indictment of the presumed injury liability of seizure-prone people.[84] Suggesting instead that such persons were "safe," it concluded that "no undue alarm need be felt over the accident possibility of an epileptic person in school or at work."[85]

Increasingly emphatic about the relative safety of seizures, school administrators further assigned a new kind of stoicism to pupils. Rather than catastrophic or disturbing events, seizures became discrete, manageable, and even mundane occurrences found in otherwise strong and healthy bodies. This mirrored a growing belief among some epilepsy specialists that "attacks should arouse little excitement," only minimal intervention, and indeed little concern on the part of those experiencing them.[86] Describing one of the many false perceptions about seizures, administrators explained that "some people have the idea that all children sleep after having a seizure. In the experience of the clinic, this is not true. Many children having epileptic convulsions get right up afterwards and go on as if nothing happened."[87] In addition to bolstering the strength and future employability of students, such attitudes ran counter to a new strain of alarm about disabilities resulting from physical injury more generally in this era. Postwar media reports in particular warned parents about the rising perils of "preventable falls" as a source of death and disability—including the possibility that such accidents could cause traumatic epilepsy lasting a lifetime.[88] For children with pre-existing epilepsy, however, the school assured families that the perception of danger surrounding seizures themselves was largely unfounded.

By accepting seizures as a reasonable risk, the school also reimagined the physical capabilities of its students. According to new concepts, the bodies of students became hardy and resilient rather than weak and sensitive to upset. In 1947, heeding advice by specialists to keep the epileptic child "busy and his body active," the school instituted its first physical education program.[89] It was led by one of the school's only male teachers, occurring in step with a broader expansion of student curriculum and clubs.[90] Beyond simply accepting the "regular pressures" of daily life, physical training threw students into motion in ways previously understood to aggravate epilepsy. Prior to this time, for instance, pupils were not allowed to play sports for much the same reason that rigorous academics were discouraged: a fear that "strenuous activity might precipitate seizures."[91] With new research, however, the school came to defend physical exercise and exertion. "It was commonly believed," a 1953 report reflected, "that the neuromuscular skill of the epileptic was below that of the non-epileptic." Following the success of its gym program, however, it concluded that the "only reason why epileptics apparently had subnormal neuromuscular coordination" was because they "never tried to learn these various skills."[92]

Within a few years of incorporating physical education courses, the school progressed to organized sports, including participation in local intramural touch football, basketball, and softball leagues.[93] Girls, because of social norms and the lesser emphasis placed on their physicality and athleticism, were not initially included and had no teams of their own until 1951. For male students, whose gender and disability seemed especially incommensurate after the war, however, sports proved transformative in the claims that it allowed staff to make about their courage, strength, and stamina.[94] Noting the boys' lack of game winning in 1947, their coach nevertheless celebrated that "in good sportsmanship, in good fellowship, and in being good losers they are unexcelled."[95] In addition to building character and cooperation, sports promoted a "competitive spirit, the lack of fear of bodily contact, and the acceptance of decisions in a sportsmanlike manner."[96] Of the athletic prowess and resilience of their "good losers" on the touch football team, administrators similarly boasted that the games continued through all conditions and weather "until snow covered the ground."[97]

Postwar students themselves described the personal benefits that resulted from their new participation in sports at the White School. Although quoted selectively, a series of testimonials in the 1949 annual report to the school board reflected enthusiasm among boy athletes. One student, Harold, claimed that sports were a welcome and useful distraction: "Sports keeps me from thinking about my seizures. The more I play the fewer seizures I have."[98] Don similarly reasoned: "My mind is better occupied. I do not worry so much about my seizures."[99] Boys also noted the physical conditioning and improved self-esteem that came with training. According to Tom: "My reflexes are better, I can run faster, I am more alert."[100] Another, Bob, remarked, "My coordination has improved, my endurance is better," while Paul stated with satisfaction that, "play has made me faster."[101] Although some boys reportedly "froze up" with intimidation during games with other schools, staff highlighted their freedom in playing and reported that, notably, "during this entire time a hard seizure has never occurred."[102]

The expansion of students' social experiences in this period also promoted the idea that they were reliable, self-contained, and indeed, "safe." Starting in 1946, children were for the first time taken off school grounds for field trips. In formal reports that year, administrators noted, "For years the epileptic children took no trips. This year the taboo was lifted with reassuring results."[103] The school made the city its classroom. Second and third graders did a unit on aviation in which they visited the airport, walked through an airplane, and later constructed a model airport in their classroom.[104] Fifth and six graders studying Detroit's industries painted a local map on the floor of their classroom and made miniatures of landmark buildings following their outings.[105] With each passing year, the school's "many trips" became more elaborate and went further afield, with excursions to sites visited by their peers in regular schools. These included parks, art museums, Ford factories in Dearborn, Michigan, and the Civic Light Opera.[106] Through participation in such activities, students both experienced and were asked to imagine a wider world than that of their homes and elementary school. The practice also challenged existing restrictions about where and under what circumstances children with epilepsy could venture into unfamiliar settings.

Ultimately, field trips, like team sports, seemed to suggest to administrators that students could broaden their activities and social horizons without having more seizures, or any seizures at all. Four years after field trips began, in fact, teachers boasted that only one "hard seizure" had ever occurred off school grounds—and even this was thought to have been "avoidable."[107] As was the case at sporting events, and indeed, as was essential to the school's image of success after the war, educators assured that "convulsions do not occur on these trips."[108] In a statement to the school board in 1951, Principal Carlotta Teagan justified a break with former policies on these grounds. "In the past," she explained, "the children from the White Special School were never taken on trips because it was believed that seizures would probably result. Stimulation and emotional upset resulting in seizures was considered to be a natural result of such excursions. . . . However, . . . if the patient is interested, active, and attentive, all disorder tends to be suppressed. . . . For the epileptic, . . . activity is good therapy."[109]

Students at the White School were also gradually encouraged to venture into public spaces alone. Citing Dilantin, as well as later anticonvulsant drugs like Mesantoin and the never marketed "AC 186" that continued to be tested at the school, administrators claimed that medications enabled this change.[110] In 1948, for instance, one student who had "never been allowed to walk to the corner drugstore alone" began "walking the five blocks to school at the request of her mother" after an agreeable uptake of Dilantin.[111] The coach service, though it persisted, did so in theory only because of the dispersion of students throughout the city and its suburbs rather than to protect children from the dangers of walking without supervision. By the 1950s, students once barred from attending spelling bee championships were supported as far as their talents took them because they were on an effective drug regimen. In such instances the school rejoiced: "One more child is able to participate as a non-handicapped individual."[112]

The emphasis on pharmaceutical seizure control again, however, misrepresented the nature and source of these changes. Although many children experienced a sizeable reduction in the number of their seizures with new medications, seizures also persisted at the school in comparable numbers before and after the war. For instance, during the 1951–1952 academic year alone, 860 convulsions were seen at the school and on its coaches.[113] Although administrators continued to admit students to the school whose epilepsy was not yet adequately controlled by medication or other means, the fact remained that seizures were neither eliminated nor even particularly reduced compared with earlier years. The primary change after 1946 then was not the experience of epilepsy as a condition characterized by intermittent seizures but rather a willingness on the part of teachers and staff to risk that seizures might occur—or indeed, to deny their presence. Such risks were warranted, it seemed, because of the perceived control and capabilities of children with epilepsy and all they stood to gain by expanding their social roles and environment.

Safe Citizens

With a growing overlap between the activities of children with epilepsy and their counterparts in regular schools, the White School's goals evolved further by the 1950s. Whereas early mission statements had emphasized the differences and limitations of its students, those of the 1950s insisted upon their seamless ability to conform. In 1941, for instance, staff had aimed to help students "realize they are different from others, to accept these facts and attempt to build a future in spite of them."[114] By 1955, however, the vice principal denied and derided such differences, arguing that "no child wants to feel different . . . most boys insist on wearing blue jeans instead of slacks just because everyone else does it."[115] Previous objectives of "acceptance" and "dignity" thus turned more conformist and assimilationist in nature. The school's role, accordingly, became more transitional; by the mid-1950s, students attended the school for a targeted six to eighteen months rather than for several years or the duration of one's education, as was once projected. During that shorter time frame, the aim was to control students' seizures, instill confidence, build skills, and either return or introduce them to the regular school system. No longer content to merely provide a second place beyond the home, the school sought to both rehabilitate and mainstream students in the fullest sense, and as quickly as possible.[116]

By the 1950s, staff also tended to discuss seizure-prone children in terms of their "normality." A pervasive concept in postwar psychology and culture, "normalcy" typically reflected white, patriarchal, and ableist ideals, emphasizing self-supportive labor and the appearance of total "health."[117] As a report to the school board explained in 1949, "Each child is expected to take his place in society. . . . Every student realizes that his education is important since he expects to be a job holder and no student expects to sit at home. The boys and girls today are considered normal except at such time when they are actually suffering a seizure and they are treated as normal children."[118] The latter comment about seizures echoed the medical and popular literature of the period, which increasingly characterized people with epilepsy as "just like anybody else" and "physically and mentally normal, save at the time of seizure."[119] Administrators, in fact, reasoned that "surely a student who has three convulsions in a year and as a result is unable to pursue his regular duties for a total of forty-five minutes during that year could not be considered a sick child."[120] Deemed "no different than a broken arm that has healed," most students whose seizures were controlled were "normal enough," at least enough of the time, to finish school, find employment, and "participate as a non-handicapped" individuals.[121] More than simply language, these ideas became standards of comportment, identity, and, indeed, personal responsibility at the school and beyond.

Such discussions also helped define paths to adulthood for students at the White School. Seeking to place their students on equal footing with boys and girls at regular schools, staff directed students toward gainful employment and marriage as long-term rehabilitation goals. Reflecting on the encouraging outcomes among its graduates in a follow-up survey in 1951, Principal Hazel reported that "each year more of them are competing on an equal basis with non-epileptics. . . . We believe that a handicapped person cannot be a good citizen until he has become a working member of his society. . . . No boys or girls want or expect others to support them in later life."[122] A questionnaire put to graduates the following year found that most young men were single and working while many young women were married and "being cared for."[123] Such reports mirrored parallel magazine articles in which men with epilepsy who "recovered from their handicap" became "self-supporting members of society" and women were "happily married with children."[124] With satisfaction, Principal Hazel concluded, "These people appear to be living normal or near normal lives as useful members of society rather than becoming burdens upon it."[125] Being "useful" and "near normal," of course, could mean either economically self-sufficient or appropriately dependent, depending on the student's gender.

Before students ever graduated or became adults, however, the uplifting effects of the school's work were said to be evident across various aspects of its program. Testimonies by boys on the 1950 baseball team, for instance, expressed novel feelings of belonging: "I feel like I was wanted," "Playing on a team has made me feel more like people who do not have seizures," and "I feel as though I am living like others."[126] Increased physical and other expectations had produced "boys of normal dreams and ambition," in contrast to the "unexcelled losers" of earlier years.[127] Academic expectations were higher too. Students by the late 1940s and 1950s were encouraged both to attend and graduate from high school and, if they could gain admission and afford the expense, to pursue college. Likewise, as staff embraced the idea that "our boys and girls are normal," the general attitude among parents was one of growing hopes for their children's futures. By the 1950s, the school's mission was thus "a single goal, namely: controlling each child's seizures, so that he can return to his neighborhood school as soon as possible."[128]

Nevertheless, visions of social belonging at the school were far from inclusive. Persistent forms of segregation and inequality meant that the "neighborhood schools" to which students returned would often differ for Black and white children and for those of differing social classes. More than separate and unequal, this also meant that African American and other minority students were increasingly sidelined from the school's broader rehabilitative aims. Whereas prewar efforts to shelter and contain seizure-prone children were somewhat universal in their focus insofar as they brought a diverse group of children together, newer goals of eliminating the appearance of epilepsy and its attendant disadvantages clearly prioritized white, American-born children with no further disabilities. According to definitions of normalcy and social citizenship that also presupposed these markers—and which hinged upon a child's ability to hide his or her relative differences—Black students and those with multiple disabilities, for example, were all but excluded. Although contradictions between the country's values of democracy and its treatment of Black Americans came into starker focus after the war, African American children with disabilities remained multiply disadvantaged in this respect. Opportunities for work, education, and professional training—however ambitious for any child with epilepsy—were often simply unavailable for others, even when rid of seizures.

Race also disappears from the formal records of the White School by the 1950s. Though the percentage of African American students only grew after World War II—with Black children constituting 21 percent of the student body in 1945 and 31 percent by the 1948–1949 school year—the school stopped recording or exploring this data after 1949 (while continuing to note, for instance, demographics of gender).[129] This shift coincided almost precisely with the shift in the school's aims from care and containment to integration—a change that also likely favored, or at least enunciated, divergent trajectories between Black and white students. During the final year that the school documented race, its annual report also remarked uncharacteristically upon the differences between the families of its African American and white students. Where the school had typically noted instances of commonality and cooperation, white parents described in 1949 were said to be more inclined to keep their child's condition "hidden," to feel stigmatized as the school's coach approached their houses, and to seek medical treatment. African American and some immigrant parents, by contrast, purportedly did not hide or experience the same degree of shame around epilepsy.[130] Thus, just as administrators adopted norms that produced differing athletic and life-planning measures for boys and girls, they assumed separate metrics along lines of race and ethnicity while rationalizing what were probably increasingly inequitable aims. Implicit was that white children had more to gain by bringing their disability "under control" as well as more to lose within their families and communities by appearing to have disability.

Despite these differences—or perhaps again highlighting them—the rising focus on mainstreaming students to regular schools and rehabilitating them to "normal life" was not, as educators implied, about making epilepsy more visible or present in Detroit's many schools and communities. Although staff and parents raised awareness through outreach work and promoted the banishment of euphemisms like "blackout" and "spell," it was epilepsy as a public education issue, and not seizure-prone children or adults themselves, that found a spotlight in the broader discourse after the war. The goal of making seizure-prone children "safe" was to fit them into society as it was, not the other way around.[131] For some school-age children who were more easily treated and possessed the right social markers, removing the trappings of safety devices and other measures meant to mitigate seizures seemed to afford them—nearly—the "non-handicapped" status to which the school aspired.[132]

New Order

In June 1946, just a few months after the protective rubber was removed from its premises, the White Special School was featured in an article on epilepsy in *Life* magazine. Toward the end of the article, in a brief description and three black-and-white photographs, the school stood as a portrait of earlier methods of management. Children queued to collect their medicine, others played in the schoolyard under the close watch of attendants, and a pair of pig-tailed girls linked arms while ascending an intact rubberized staircase. Like the medical wonders of EEG and Dilantin that populated the preceding pages, such images suggested order and treatability. Where epilepsy was not helped by emerging medical developments, it was contained within a friendly institutional setting.[133]

One of the three photos of the Detroit school and its pupils featured in *Life* magazine, June 3, 1946. Photo by Mark Kauffman. Mark Kauffman/The LIFE Picture Collection/Shutterstock

The presentation of the White School in *Life*, however, was somewhat outdated by the time of the article's publication in 1946. In particular, it belied recent school policies that emphasized lives for seizure-prone children beyond their homes, special schools, or the watchful care of others. By the early postwar era, students were increasingly expected to transition to regular schools, go to high school and college, perhaps get married and have children, and find stable employment. Facilitating this outlook were emphatic assertions among staff and parents that students were "normal" except for seizures. At the same time, seizure-prone children did not become suddenly "safe" in the ways generally depicted. Such efforts occurred in the context of greater though imperfect pharmaceutical control as educators endeavored to minimize epilepsy's effects. New medications made occasional convulsions and losses of consciousness seem more remote and manageable, making it possible to present students' newfound independence as an outgrowth of their personal control.

Removing prewar measures of protection also required protective measures of new kinds. Newly responsible for their physical self-mastery, seizure-prone children had to be taught, for instance, to unlink arms, to endure scraped knees and other injuries, and to take losses on the playing field in stride. Unlike preceding classes, they also had to contemplate new sorts of adulthood. The breakdown of seizure control, on field trips for instance, no longer represented a failure on the part of the institution to properly insulate the child. Instead, seizures more often became a sign of one's inability to adhere to treatment or properly self-regulate. Removing protection, then, was not only about taking risk; it was also part of the therapy. Seemingly ordinary activities such as playing in school tournaments and walking home alone became the very practices by which new standards of belonging might be achieved—and they fell almost exclusively to students as they navigated these new roles.

The White Special School, as an institution solely for children with epilepsy, closed in 1956. Anticonvulsant medications could be dispensed at home, rendering the school, as originally conceptualized, obsolete within two decades of its founding. Yet far from the rehabilitative success story it projected, the school in some ways mainstreamed one group of children and left behind others whose constellation of disabilities disqualified them from its vision of integration. After 1956, it became the White School for Crippled Children, serving those with multiple disabilities, including epilepsy.[134] Simultaneously, students who most clearly embodied the school's rehabilitative goals met new challenges. With rising expectations for control but no cure, all children with epilepsy, as before, lived with seizures and seizure potential, social intolerance, and possible exposure of their disability. Casting off the physically protective atmosphere of the school, a new generation of seizure-prone graduates instead found a precarious safety rooted in the promise and appearance of their control and self-sufficiency.

4

Mobility and Motor Control

Epilepsy in the Age of the Automobile

Unfolding in step with the changes happening around seizure-prone children in pre– and post–World War II Detroit were those occurring on America's burgeoning roadways. In the summer of 1939, a salesman from Kansas City had a particularly noteworthy seizure while driving across the recently constructed San Francisco–Oakland Bay Bridge. In the dramatic and much recounted scene, the man's speeding car careened for several minutes between lanes of traffic before pursuant police cars ran it into a guardrail. Although no bodily harm came of the incident, newspapers highlighted the spectacular danger of the event. Apprehending authorities, reports stated, had found the man still in "spasms," his foot clamped down on the accelerator with the engine racing. In a single moment, the otherwise unassuming businessman, who had driven his car all the way from Missouri without incident, became a symbol of the dire threat that epilepsy posed to public safety.[1]

The case of the Kansas City salesman, in its evocation of twisted metal and bodies at great speeds and heights, seemed to reveal a hidden but significant danger. Exposed in one man's seizure was not only the unseen peril of the haphazardly stricken person but the tenuousness of road safety itself. Shortly following the accident, the state of California designated epilepsy as a reportable disease, a rule that required doctors to submit the names of epileptic patients to local health and motor vehicle authorities.[2] The measure resulted in the revocation of thousands of California licenses and, by the end of World War II, was adopted by several states across the nation.[3] Such laws, however, did not go uncontested. Soon after the war, some Wisconsin doctors and disenfranchised drivers began to challenge their own recently established reporting laws, arguing that certain people with epilepsy were safe to drive. By 1949, the state became the first of several to selectively temper its policies, certifying some seizure-prone people as "safe driving risks" and granting them licenses on a limited basis through a process of medical review.[4] In a single decade then—much like at the White Special School—the "epileptic driver" was defined as a social problem and just as quickly reclassified as a candidate for the driver's seat.

This chapter explores how seizure-prone people went from being treated as dangerous drivers to safe drivers between 1939 and 1956. In particular, it shows that the car and the public road were key sites in which questions of the safety—and by extension, the rights and responsibilities—of seizure-prone people were more generally remade. More than any other cultural space or institution, a driver's license demonstrated that those with epilepsy could be "regular" citizens. In fact, before the right to marriage or equal education, driving became the first civil privilege won by people with epilepsy. A license showed that one was not only a competent driver but also a safe and controlled person, worthy of inclusion in the postwar social order.[5] Equally, however, the chapter suggests that questions of control were far from straightforward. Although advocates of Wisconsin's limited licensing laws after World War II placed their faith in the seizure-controlling properties of anticonvulsant drugs, driving remained a high-risk activity and no medication guaranteed seizure cessation.[6] Indeed, even in the best medically controlled cases, people continued to experience seizures, or at least the possibility of them. Thus, at a time when the rules of the road were formally taking shape, licensing "epileptic drivers" became an exercise in deliberating on new parameters of risk and erecting the appearance of control and safety where they had not previously existed.

To understand how definitions of medical control influenced those of road safety at a critical juncture in the histories of both epilepsy and the American car, the chapter focuses on two main case studies: California and Wisconsin. Through analysis of each state's landmark laws for drivers with epilepsy, it reveals how two radically different models of driver safety evolved. States that initially followed California's example in 1939 safeguarded public roads by exposing people with epilepsy and denying them driving privileges in a manner experts described as "akin to quarantine."[7] Conversely, after 1949, Wisconsin's converts found safety in partial and probationary licensing practices, whereby appointed medical boards assessed drivers according to their "reliability" as patients and citizens—often beyond their actual seizure record. Moreover, while responsibility for control fell to doctors and local officials under reporting legislation in 1939, by the mid-1950s it increasingly resided with seizure-prone drivers—who, in accepting a "limited license," were deemed "controlled epileptics" in command of their treatment, their own decisions and actions, and the welfare of themselves and others on the road.

Tracing new approaches to seizure-prone drivers mid-century, the chapter engages with several literatures at the intersection of public health and safety history. Most broadly, it evokes a critical historiography on the management of infectious diseases. Although the chronic nature of epilepsy distinguishes its case, those appointed to determine the road readiness of seizure-prone applicants made evaluations similar to those made by public health officials about quarantines, blood-banking protocols, and immigration bans related to various infectious diseases.[8] Defining "high-risk" persons and reasonable levels of risk, both physician advocates of driving bans and the medical boards that eventually approved seizure-prone drivers also fundamentally understood their work as a form of disease control aimed at protecting public health and safety. The case thus provides a novel illness perspective and extra-clinical vantage point from which to consider classic themes within these histories, including questions of state and medical power, tensions between individual liberty and public welfare, and—most pertinently here—restrictions on an individual's mobility in the broadest sense.[9]

This case also contributes to more recent work on the medicalization of the automobile and its drivers. As historians have shown, rising expert involvement characterized a number of debates and interventions surrounding speeding, drunk driving, and "accident-prone" motorists as early as the 1920s.[10] Less well documented is the parallel role that physicians and psychologists played as governments established nascent driver's licensing laws and criteria for issuing permits.[11] At a time before most states required in-person applications or testing, the idea that people with "medical defects" could sidestep authorities and obtain licenses was a growing source of concern.[12] Drivers with epilepsy became an early example of how physicians and lawmakers defined "driver fitness" and managed fears about temporary impairments ranging from heart attacks and strokes to diabetic episodes.[13] Whereas initial concern throughout the interwar decades had focused on people with poor eyesight, mental illnesses, and low intelligence—each of which was presumed to be permanent, consistently experienced, and thus easily detected—by the World War II and postwar era, intermittent and unpredictable states of impairment seemed to pose a more insidious threat. The chapter reveals the ways in which notions of safety and control were negotiated vis-à-vis seizure-prone people mid-century and how these categories in turn came to define experiences of epilepsy.

Finally, in examining disabled or "medically impaired" drivers, the chapter revisits twin conceits within cultural histories of the American car: first, that by the 1950s driving was a core signifier of civic belonging, and second, that responsibility for road safety was increasingly assigned to individual drivers over states and manufacturers.[14] Permission to drive could indeed bolster public confidence in the mainstream capabilities of people with disabilities, as argued of the bans lifted on deaf drivers in the 1930s.[15] Public anxieties about the driving competency of female and teenaged drivers by the 1940s and 1950s correspondingly tended to reflect concerns about their expanding mobility and social roles.[16] Yet, whereas promoters of women and deaf drivers emphasized each group's supposed positive attributes, such as "patience" or "superior caution," licenses for people with epilepsy alone came with expectations of absolute and inconspicuous control. Seizure-prone drivers also engendered an unparalleled level of mistrust once licensed. Thus, rather than simply liberalizing laws and extending civil rights as suggested by physician advocates, the licensing of epileptic drivers also increased the restrictions and responsibilities placed upon them. Ultimately, granting licenses would prove safest—not because it reflected trust in the control afforded by new medications, but because it addressed an enduring fear of people who concealed their seizures and might crash unexpectedly. Where matters of road safety were concerned, epilepsy's invisibility in the public eye remained its greatest danger. Shouldering the burden of this construction of safety—and indeed, responsibility for the limits of control—was the newly licensed epileptic driver.

1939: Reporting Epilepsy

According to physicians at the forefront of California's reporting laws in 1939, the car and the person with epilepsy were wholly incompatible. "With the increase of mechanization in this century," argued doctors W. E. Carter and Richard Harvey at the University of California, San Francisco, "epilepsy has taken on a new significance in the social order. Before the time of the automobile the epileptic was seldom a menace to others and infrequently injured himself. Now that everyone is an automobile driver, real or potential, the situation is vastly changed."[17] Placing this dangerous confluence along a longer continuum of technological progress and peril, Carter further explained, "That every advance in the industrial world carries a new health problem is nowhere better exemplified than in the relation of the convulsive state to moving machinery. . . . Recent experiences all over the country indicate that persons who are subject to episodes of unconsciousness are an increasing menace on the highways."[18] Insofar as drivers with epilepsy created a novel "health problem," they also required physicians' attention. Those most vociferous about curbing such drivers, in fact, argued that it had "become incumbent upon practitioners of medicine" to protect both the "victim of a convulsive state" and "others who might be the victims of his acts."[19] No longer a mere patient or private citizen, the seizure-prone person who drove a car was hazardous to the public's health and safety.

New anxieties about drivers with epilepsy following the Bay Bridge crash coincided with broader concerns about road safety and reckless driving in the years preceding World War II.[20] Automobile ownership grew exponentially throughout this period, roadways expanded, and the United States was increasingly "a nation that travels on wheels."[21] Many Americans also became acutely aware of the injury and death toll created by cars. Automobile-related accidents and fatalities rose each year throughout the 1920s and 1930s, culminating in public outrage and the organization of numerous safety councils across the nation.[22] As historians also note of this era of increased "crash consciousness," lawmakers often placed blame for accidents with individual drivers rather than with the technical aspects of the car or the condition of the roads.[23] As driving became more firmly entrenched in American life, such parties instead deemed the "human factor" in driving—often positioned in opposition to the technological features of the car—to be an inevitable source of "failure," needing various means of external regulation and control.[24]

State and federal regulations emerged throughout the interwar decades to set standards for speed limits, street lighting, signage, and safety specifications for car headlamps and mirrors.[25] Driver safety itself, however, fell largely to medical professionals. The "unconscious motorist," in fact, became one of many "medico-traffic problems" of the time, in which consulting physicians considered the impacts of human health and impairment within and beyond the boundaries of the car.[26] By the late 1920s and 1930s, doctors were closely involved in the design of safety belts, the earliest crash and breathalyzer tests, and new treatments for injuries related to car accidents.[27] Those not directly involved in such projects also voiced their concerns. Many neurosurgeons, for instance, complained that car crash victims flooded their operating rooms and impeded the work of operating on patients with preexisting conditions, while neurologists noted that traumatic epilepsy in particular would be exacerbated by the country's rising car centrism. Boston neurologist William Lennox, for example, warned that "as vehicles inexorably increase in number and become faster, more powerful, and more arrogant, seizure-begetting head injuries increase."[28] For "public-spirited" physicians of the period, the car was a "potentially lethal instrument" that many felt morally compelled to mitigate and that, for some, extended their professional reach.[29]

Concerns about epilepsy in the context of driving were nevertheless unique and went beyond previous medical preoccupations about the condition. What particularly worried California doctors and public health officials by the late interwar period were not the usual questions of the epileptic's biological fitness. Such eugenic concerns had informed debates on issues ranging from marriage to military preparedness throughout the early twentieth century.[30] Rather, commenters following the Bay Bridge crash focused primarily on the potentially devastating effects that even temporary losses of consciousness might cause at the wheel of a car. According to one physician, "the operation of a motor vehicle under the trying conditions of highway traffic today demands that the driver exercise constant vigilance and care."[31] Seizures, in their unanticipated interruptions to consciousness, were thus by definition untenable. As Drs. Carter and Harvey described, "The nature of the attack, coming as it does without warning in most cases, renders it particularly dangerous."[32] Emphasizing the peril of even a single seizure, their Detroit colleague Dr. L. E. Himler echoed, "As a road hazard the danger of epilepsy lies not in the frequency of its occurrence, but rather, like lightning, in its suddenness and unpredictability."[33] Still others underscored the modernity of the dilemma and the special case of the car, noting that whereas horse-drawn carriages forgave the "holder of the reins" a momentary lapse in consciousness, the "gasoline-driven vehicle" was a "particular foe of the epileptic."[34]

In addition to accentuating epilepsy's perceived hazards, the new attention-based demands of the car also expanded the category of those with epilepsy who were considered unsafe. Historically, grand mal convulsions and the people who experienced them had been most feared. Petit mal seizures—those which might be no more disruptive or noticeable than a momentary disengagement from conversation—seemed less troublesome. With driving, however, such "minor" and less detectable seizures became equally if not more threatening. As Carter and Harvey argued in 1942, "The question of petit mal is one that must be considered in reviewing this problem," since, they explained, even a short lapse in consciousness could be a matter of life and death.[35] Others went further to suggest that sufferers of petit mal seizures were especially dangerous since their condition was more likely to go undetected by physicians and those around them.

After 1939, people who experienced seizures of any kind and frequency in the United States were increasingly deemed bad road risks.[36] Following Norway's precedential decision to make epilepsy a reportable disease in the spring of that year, physicians in the state of California first lobbied for a similar law.[37] In addition to the problems seemingly inherent in epilepsy, doctors also argued that their car-centric state was particularly vulnerable given its ballooning population, vast territories, extensive highway system, and ever-increasing ratio of automobiles per capita.[38] The bill to effectively ban those with epilepsy from driving passed rapidly in September, culminating in a statute that required physicians to report all known people with epilepsy to the State Department of Public Health. That agency in turn was required to report the names of those identified to the state's Motor Vehicle Department in order to revoke or deny future licenses.[39] While many states by this time possessed generic restrictions against individuals deemed "unfit to drive," California became the first to formally prevent those with epilepsy from holding a driver's license. Previous rules, insofar as they specified exclusions, had usually listed low intelligence, insanity, and poor vision among the restrictions. California's reporting law, however, was the first of such exclusions to be written into law and moreover to require physicians, upon threat of misdemeanor, to report their patients to bureaus of state government.[40]

Touted as a "quarantine law" among its proponents, compulsory reporting transformed epilepsy from solely an individual or family illness to a disease firmly within the bounds of public health. Lifelong but also now immediate in its dangers, epilepsy seemed to require interventions similar to those exercised in cases of communicable disease. "Excluding acute infectious processes," Carter and Harvey offered by way of comparison, "perhaps no disease bears a greater hazard than does epilepsy."[41] Supporters also linked epilepsy's prevalence to that of diseases like tuberculosis and syphilis as a means of both characterizing and quantifying its sudden and far-reaching threats.[42] Adopting the analogy of infection and containment, doctors presented new practices of reporting people with epilepsy and revoking driver's licenses as a type of community "preventative medicine" that curtailed citizens' personal freedoms to serve the greatest good.[43] Public roads and highways were also perhaps no great stretch from the regulation of epilepsy that occurred at borders, in the military, and elsewhere. The suspension of driving privileges in this manner nevertheless extended epilepsy's monitoring to new spaces and levels of formality.

California's compulsory reporting laws also reflected a more general awareness and alarm surrounding epilepsy by the late interwar and World War II period. During World War I, medical screening and concerns about the fitness of American soldiers had increased epilepsy's otherwise minimal institutional and public visibility, with results projecting that roughly one in every hundred Americans was "epileptic."[44] However, whereas such studies had illuminated the statistical prevalence of epilepsy, concerns by the Second World War focused less on occurrence than on the unseen place—and movement—of sufferers in the American public. As Detroit-based physician Dr. Himler noted of the preponderance of people with epilepsy living seemingly "normal" lives in 1941, "The necessity of effective measures to eliminate the danger of epilepsy in traffic is self-evident when it is recalled that of the 500,000 or more patients with epilepsy in the United States, fully 450,000 are not in institutions but are living in the community."[45] More alarming, he continued, was that "a considerable portion of patients with epilepsy do not refrain entirely from driving."[46] Harvey and Carter similarly argued that many of their outpatients frequently "came to the clinic driving dilapidated cars, sometimes from points a hundred miles away."[47] What troubled doctors most was that beyond such moments of visibility, the lives of their patients remained completely opaque, including the actual number who possessed licenses, their driving habits, or, indeed, the accident record perhaps suggested by their "dilapidated" vehicles. "One can only speculate," Dr. Himler concluded, "on what proportion of the many accidents reported as driving on wrong side of road, driving off roadway, and reckless driving might be related to epileptiform states."[48] The few known and reported accidents caused by confirmed seizures, in other words, might be indicative of a much larger, and mostly unseen, problem.

At the root of concerns about epilepsy's low visibility in this period was a desire to identify people with epilepsy despite what some described as its tendency to leave little discernible trace upon sufferers. As one physician noted of epilepsy's "ambiguous physical standards" and implications for a driving society, "The licensing agency has no means of ascertaining such a disabling condition as epilepsy unless by some chance it learns of an accident while the driver was in an epileptic fit."[49] In addition to epilepsy's inconspicuous and latent nature, many physicians also assumed that people with epilepsy rarely disclosed their condition. According to Carter and Harvey, "Patients who have epilepsy will deny the presence of the disorder, even to the point of swearing falsely in an application for a driver's license."[50] Condemning this as "perjury," proponents of reporting presented such persons as untrustworthy and lawless, inflated by their success in hoodwinking authorities. "The average so-called epileptic," Carter wrote, "often refuses to acknowledge the gravity of his condition, and drives an automobile with impunity. He is not permitted to operate a train, a bus nor a streetcar; but when he seats himself at the controls of an automobile, his ego is expanded, he has no one to say nay, and the results are often tragic."[51]

Yet even in cases where people were not suspected of willfully concealing their epilepsy, its imprecise definition and symptomology was said to foster a culture of "confusion and omissions."[52] As in employment and other high-stakes scenarios, doctors cited stigma and an "attitude of concealment" as the cause of patients' tendency to knowingly or unknowingly bear false testimony on driver's license applications.[53] People who had never received a diagnosis or were unclear about the parameters of the illness—those who, for instance, believed that epilepsy meant convulsions and not intermittent petit mal seizures—might answer in the negative and obtain a license.[54] Those individuals caught driving, however, were almost always depicted as inadequate judges of their disabilities. As explained one man whose license was revoked by Detroit police in 1942 after he had driven an estimated 18,000 miles, "I knew that I would be rejected if I told the truth. . . . I didn't think anything of the spells at that time; I knew I had epilepsy but thought I was O.K. to drive."[55] While authorities and physicians attributed the man's testimony to ignorance and evasion, some drivers probably genuinely remained uncertain of their symptoms or estimated that the particular presentation of their seizures would not impede their ability to drive.

With the United States' entry into World War II, epilepsy's amorphous threats only seemed to increase. Although car-related accidents initially diminished during the war owing to tire and gasoline shortages and fewer drivers on the road, accidents and fatalities rose significantly after 1941. Critics argued that such injuries and deaths rivaled those of the battlefield.[56] As a special Medical Committee on Motor Vehicle Accidents anticipated of domestic road conditions shortly after Pearl Harbor: "Motorists will be driving on poorer tires, new cars will not be available, and every jalopy that can navigate may be put to use."[57] The committee also ventured that "the general mental and nervous strain which is always present in a time of national emergency" would be "especially prevalent" with war, leading to more distraction and a greater number of accidents among drivers.[58] In addition to stress and unsound vehicles, the group noted the growing movement of the nation's many defense workers, reasoning that "as more younger and able bodied men enter the armed forces, more civilian duties will be taken over by older, less alert men and women."[59] Though the committee did not specify that many of these jobs would be filled by workers with a variety of disabilities, their anxiety about the quality and safety of "less alert" and ostensibly less capable commuters was clear as the distribution of people and work shifted across the country.[60]

Proponents of reporting laws for epilepsy in California, however, were comparatively explicit about the risks that typically male, and frequently migrant, out-of-state war workers posed to their state and its citizens. Some physicians and newspapers, for example, described displaced men who arrived from all over the Midwest and "sometimes as far as the Appalachians" for work, resulting in an influx of workers, and drivers, with epilepsy. As Dr. Carter reflected in 1944, "Each one seems to feel a new start can be made, that the climate will improve his health, that no one will know his history, and that the high wages paid for even the commonest of labor will improve his economic status."[61] Such accounts echoed aspects of Dr. Tracy Putnam's contemporaneous characterization of "fugitives from epilepsy," of those individuals who attempted to outrun both seizures and public knowledge of their condition. They also evoked concerns about itinerant men during the Great Depression, as people with epilepsy who traveled across state lines, desperate for work, seemed to present similar potential dangers.[62] Only "the medical profession," the Medical Committee on Motor Vehicle Accidents argued, "can aid by keeping the public cognizant of these facts and by re-iterating the need for maintaining high physical standards in our civilian population."[63]

As the imagined demographics of drivers continued to change, support for compulsory reporting grew. Informal reporting already existed in cities like Detroit, where civilian informants disclosed the names of people with epilepsy to city courts and police departments. Advocates of compulsory re-porting, however, argued that a better and more comprehensive system was needed to discern epilepsy among the country's many drivers. This placed responsibility for detection and public safety squarely with physicians.[64] As Dr. Carter put it, reporting by family doctors and specialists was essential to determine "whether or not the holder of a driver's license was safe to be on public streets."[65] Knowledge of "special diagnostic procedures utilized in the study of epilepsy" made neurologists, including numerous supporters of driving bans, especially well equipped for "valid judgment."[66] Technologies like electroencephalography, for instance, offered them exclusively the one "true" and only "objective record."[67] Though voluntary reporting by neigh-bors and coworkers could expose unsafe drivers within communities, man-dated reporting by medical practitioners with superior tools and training helped ensure that even subtle and "more serious cases" of epilepsy would be accounted for.[68]

During the first six months that California's reporting law was in effect, 2,780 cases were reported. Less than five years later, there were a total of 8,696 cases, with licenses revoked at a rate of three per day.[69] Advocates of the law, though pleased with their progress, understood this to represent only a third of afflicted drivers in the state. Articles in medical journals con-tinued to emphasize the public danger of epilepsy as well as the joint role that doctors and public officials played in tempering impaired driving. In 1946, the American Association of Motor Vehicle Administrators (AAMVA) gave reporting legislation a boost by voting in favor of the measure at its annual meeting. Shortly thereafter, reporting laws were adopted in Iowa, Kansas, Oregon, Indiana, New Jersey, Wisconsin, and elsewhere.[70] Califor-nia's law, in fact, continued to gain favor and momentum throughout the late World War II years and into the postwar era.[71]

1949: "Safe Risks" and Limited Licenses

As the war ended, critics of reporting laws began to suggest that some seizure-prone people might be safe to drive.[72] Wisconsin's Department of Motor Vehicles, for example, though it was among those states to follow California's lead in 1947, received several complaints from physicians and people with epilepsy, just as new zeal for road safety grew in response to rising automobile use, suburbanization, and highway expansion plans.[73] Becoming a vocal spokesperson against the reporting measure in Wisconsin, Dr. Charles Crownhart, secretary of the state's Medical Society, argued that submitting the name and address of every person diagnosed with epilepsy "put the doctor on the spot" and undermined doctor-patient confidentiality.[74] Others underscored the ways in which such laws forced physicians to take permanent, prejudicial action. Milwaukee neurologist Dr. Edward Schwade, for example, protested that it was "certainly in bad taste for medical men to report the epileptic any more than . . . cases of appendicitis, gall bladder, or heart disease."[75] Moreover, since seizures did not always indicate epilepsy, the provision left little room for diagnostic ambiguity or error.[76] Opponents argued that this risk was especially concerning for people with other underlying or temporary conditions, as well as for veterans who might be incorrectly diagnosed as "epileptic."[77]

Misgivings about reporting unfolded on a wider scale as well. As early as 1946—the same year that the AAMVA announced its support of the measure—officials at California's Department of Motor Vehicles uncovered a range of dissenting opinions. When DMV official Mark Irons and Los Angeles doctor Mark Albert Glaser sent a questionnaire to American members of the International League Against Epilepsy, the results revealed experts' doubts about epilepsy's threat.[78] Although each of the 127 answering physicians agreed that epilepsy was a "definite motor vehicle danger" and that those diagnosed should not be issued permits, some believed that there were "certain conditions" under which such persons might be allowed to drive.[79] These conditions included having a licensed chaperone ride with the driver and allowing people "like housewives" who drove within a limited radius, or those who experienced "freedom from seizures" for several years, to drive.[80] Glaser and Irons concluded that the question of licensing people with epilepsy remained an "intricate problem" rife with liabilities.[81] Their findings nevertheless showed that support for bans was far from unanimous and that reporting was far from the unequivocal public good that California's early proponents touted.

In addition to doctors' own professional and diagnostic concerns, critics of compulsory reporting argued by 1947 that people with epilepsy in fact posed a minimal road hazard. Using data from a 1939 National Safety Council report, Wisconsin's advocates showed that "epileptic drivers" had accounted for only a fraction of 1 percent of known fatal traffic accidents that year.[82] While rivals would argue that this low number reflected the underground nature of the problem—and moreover only included statistics from the pre-reporting era—those in Wisconsin disregarded these claims. Admission statistics from the Milwaukee County Hospital Emergency Room in 1948, for instance, revealed that only one case of car accident injury or death that year was attributed to epilepsy. The vast remainder, Dr. Schwade and his colleagues reported, were caused by "acute alcoholism, hasty emotional outbursts, defective roads, and hazardous crossings."[83] Of the clean driving record and relative safety of people with epilepsy, they concluded, "Certainly, this segment is the least dangerous from the standpoint of driving an automobile."[84]

Epilepsy, a Public Health Problem . . . II

Drugs Allow Epileptic
to Live as Normal Man

Headline from "Epilepsy, a Public Health Problem" series, *Milwaukee Journal*, Feb. 16, 1953. © Milwaukee Journal – USA TODAY NETWORK via Imagn Images

Most notably, however, Wisconsin physicians and patients who rejected compulsory reporting argued that drivers with epilepsy could be safe on the road since so many experienced better control of their seizures with new and increasingly available medications. Following the release of Dilantin, several doctors had meditated on the relative safety of the "treated epileptic."[85] Leading expert William Lennox argued that while "no one condones the license of the active epileptic," for many, drugs and medical oversight "make them safe persons."[86] People with epilepsy who protested the law also commonly asserted that they were "treated" and made safe by medical supervision, while others argued that compulsory reporting's main flaw was that it failed to "differentiate between active and quiescent epilepsy."[87] Even California's Glaser and Irons, though they sided in favor of a "plan of non-restoration" of licenses in 1946, conceded that such a policy seemed "unfair to persons in whom an effective control has been established under therapy."[88] As distinctions between "controlled" and "uncontrolled" and between "active" and "quiescent" gained momentum after the war, they also dominated discussions of driver safety.

The safety afforded by pharmaceutical control, though subjective, stood to make those with epilepsy competent drivers and, implicitly, worthy citizens. Coinciding with wider postwar efforts to positively recast the public image of people with epilepsy, Wisconsin's campaign emphasized the "health" and "normalcy" of seizure-prone drivers, who were capable of economic self-support and control.[89] Much of its rhetoric focused on young men, given the displacement of workers during the war and efforts to rehabilitate veterans with acquired epilepsies. In addition to being a marker of social belonging and adulthood, a driver's license could be crucial for gaining and sustaining employment or attending college.[90] Emphasizing again the transformations provided by medical treatment, Dr. Schwade argued that while "the epileptic who is uncontrolled is a hazard to himself and to society," the "epileptic under control" was "an excellent worker, a good student, and perfectly normal in all other respects."[91] Control therefore not only ensured a degree of safety amenable to driving a car but also facilitated one's image as a "very useful citizen."[92]

By 1948, only a year into Wisconsin's enforced reporting procedures, physicians in the state drafted a bill in conjunction with the Motor Vehicle Department to issue driver's licenses to some seizure-prone people, supplementing rather than repealing the reporting legislation.[93] Participants like Drs. Schwade and Crownhart deliberated on matters such as how to define epilepsy, what constituted adequate seizure control, and by what means and authority licensing decisions would be made. Most notably, the group revised the stipulation "free from epilepsy" in the 1947 law to the suppler wording of "under good medical treatment and control as not to constitute a hazard while driving."[94] Early suggestions that certification should be predicated on in-person electroencephalographic testing were quickly discarded owing to the expense and perceived potential for inaccuracy. Although the group failed to convince the Motor Vehicle Department to revoke reporting altogether, by March 1949 they were successful in passing a bill to issue temporary licenses to some approved persons, renewable upon recertification by an appointed Medical Review Board.[95]

"Epileptic Hearings" and the Construction of Safety

Limited licensing in Wisconsin, like widespread compulsory reporting practices, sought to improve the safety of the country's roads and highways. Yet given the very possibility of a legitimate "epileptic driver" under the new 1949 law, levels of safety and control also had to be defined. Following the passage of Bill 85, a Medical Review Board formed to evaluate individual applicants, meeting at the Milwaukee Courthouse as soon as ten people in the state sought review. It was composed of three members: neurologists Dr. Schwade and Dr. Edward Roemer, and Department of Motor Vehicles commissioner B. L. Marcus, who together possessed the authority to grant temporary licenses.[96] During what were called "Epileptic Hearings," applicants met individually with the board, had their personal files reviewed, and underwent questioning. Successful applicants received licenses, renewable every six months, whereas those denied were instructed to reapply after a period of two or more years without seizures. Although far more applicants were initially denied than approved, within a few years roughly half of all applicants gained licenses. Of the 1,201 people reported in 1956 and the 651 who ultimately came before the board that year, for instance, 364 were approved, 212 were denied, and 75 failed to file.[97]

Officially, an applicant's safety to drive was determined according to the board's thorough review of his or her medical history, driving record, and reports from personal physicians. New to Wisconsin's law was the "Certification of Seizure Control," a document that required family physicians to swear as to a patient's period of freedom from seizures, as well as to their "remote and unlikely" recurrence under continued medical care.[98] Publicly, the Wisconsin Medical Society also endorsed limited licensing on the grounds that "medical control is possible."[99] In practice, however, board evaluation was highly subjective and based as much on perceptions of an applicant's suitability for a license—including one's knowledge of epilepsy, adherence to treatment, and driving needs—as on any formal or quantifiable measure. Through critical in-person evaluation, applicants were judged on their character, comprehension, and compliance, if not their perceived "reliability"— often beyond matters related to certified seizure control.[100]

Typical, for instance, was B.R., a filling station attendant and World War II veteran who was approved for a temporary license in 1954.[101] His records indicated that he had been seizure free since serving in the army three and a half years before. His physician had signed the appropriate certificate of control, and he had no history of traffic accidents. Additionally, the board noted that B.R. appeared to be in good physical condition and to understand his diagnosis, inferring that he "[could] be relied upon to take his medication faithfully."[102] At the end of the brief meeting, it was their unanimous opinion that he was "a proper person to be issued an operators' license."[103]

Central to gaining a license, as in B.R.'s case, was presenting as a good patient. Successful applicants had to be under the care of a physician, adhere to the prescribed treatment, and accept their diagnosis in accordance with orthodox medical interpretation. In contrast to B.R.'s "understanding" of his epilepsy and faithful ingestion of his medication, others during the same 1954 hearing were denied licenses for reasons beyond their level of seizure control. K.H., for instance, though he had a history of "episodes," was said to "not believe it is epilepsy because he does not 'thrash about.'"[104] Similarly, although I.J. had experienced a seizure within the year leading up to his review, more concerning to the board was that he "denies he has epilepsy" and refused medical care.[105] In addition to rejecting or evading diagnosis, those who discontinued medication without physician approval were often denied—even if their seizures had not returned.[106] One applicant in 1951, for instance, was rejected when he admitted he "did not follow the prescription of the doctor thoroughly."[107] Others who self-medicated with patent drugs or attempted to "gain a cure" with popular "electrical treatments" were similarly denied licenses—as were those suspected of obscuring aspects of their medical history, those who changed doctors too frequently, and those who supplied testimony inconsistent with that of their physicians.[108]

Certainly, many applicants were granted licenses because they were seizure free for the two or more years, constituting adequate "seizure control" under Bill 85. Many applicants who were not "controlled"—whether they took medication regularly or not—were rejected.[109] However, such facts alone rarely warranted, or indeed, prevented the issuance of a license. Some people were deemed "safe risks" even when they did not meet the criteria for seizure cessation. A carpenter's helper, for instance, noted to be "a reliable individual of good character and habits," was deemed "safe" after he had gone only six months without a seizure.[110] Similarly, a woman who had had a seizure six months before her review in 1951 was nevertheless issued a license after submitting to a medical exam.[111] Limited licensing thus allowed for flexible categories of control and safety—ones in which neither the history nor occurrence of a person's seizures precluded gaining a license.

In general, applicants whose seizures were found to be caused by conditions other than "true epilepsy" were more likely to have their licenses approved. Such cases included a salesman whose license was restored once he treated the anemia causing his seizures, a janitor whose license was "safely reissued" when doctors properly identified the episode he experienced as a stroke, and a truck driver whose convulsions were caused by syphilis.[112] Licenses were issued to three women whose seizures were later said to be "fainting spells" or simply "brought on by stress," as well as to those who recovered from head injuries.[113] Licenses were also restored when a misdiagnosis was confirmed—even if the applicant's seizures had not abated. Two men in 1950, for example, regained their licenses shortly after they underwent operations for brain tumors and surgeons deemed them "not epileptic."[114] A number of applicants referred by the Veterans' Administration likewise gained licenses after their wartime diagnoses of epilepsy were changed to "battle fatigue" or "concussed."[115] Whether because such etiologies were less stigmatizing or suggested more promising courses of treatment, applicants who found alternative labels fared better under board review.

Conversely, certain factors could disqualify an applicant, even when their seizures appeared to be adequately controlled. Given the importance of onsite evaluation by the board, no-shows were always denied licenses, despite the fact that some had records indicating seizure control and valid reasons for failing to attend their reviews.[116] The board justified this policy on the grounds that "without personal appearance it is impossible to evaluate reliability."[117] A history of motor vehicle accidents, most often but not always related to seizures, also disinclined the board to issue a temporary license. A seizure-free high school student, for instance, was denied a license in 1950 because he had previously struck a building while having a "spell."[118] Comparable was a farmer whose license was denied the following year because of a previous convulsion in which he fell from his tractor.[119] Some applicants were also disqualified based on their drinking habits, including a different farmer whose sporadic seizures by all accounts followed "evenings of heavy drinking."[120] A history of alcohol use, however, unlike a history of accidents, usually prompted "further investigation" rather than flat refusal. Those who showed they could abstain had their licenses restored at subsequent hearings, whereas those with a record of seizure-related crashes rarely did.[121]

Beyond an applicant's apparent character and level of compliance, however, the perceived "need" of the applicant was perhaps most crucial to gaining a license. Those with successful applications usually had clear, employment-based reasons for needing to drive. These, too, were often considered above records of seizure control. Seeing driving as essential to one's livelihood, Wisconsin's board reasoned that for men, in particular, "without use of a car, their standard of living would drop."[122] N.M., a thirty-seven-year-old chemist, for example, was approved on the basis of not only his seizure record but also his need to drive to work and for lab-related errands.[123] I.R., a twenty-six-year-old salesman and former fireman whose new job was thirty miles from his home, also received a positive outcome, as did those who said they required a car for promotion in their work.[124] One employee of an electric company needed a car to supervise transformer lines at various sites, while another man, a mason, was approved for a license to travel between construction jobs.[125] Veterans were also regularly granted licenses, both because the VA's farm training program required them and because a diagnosis of epilepsy could jeopardize rather than ensure disability payments, thus increasing one's need for steady employment.[126]

Work-related reasoning, however, did not always result in the acquisition of a license. Those who sought approval for driving-based jobs were typically denied, since people with epilepsy under the 1949 law were restricted to the operation of "pleasure vehicles only." J.B., for example, was unsuccessful in his attempts to gain a license for a "light trucking job" in 1954, as was a taxi driver with a history of blackouts, including one that resulted in a collision with a fire hydrant.[127] At the same time, a driving-intensive job did not automatically disqualify an applicant for a license—particularly if a personal car was used. At the same hearing that disqualified J.B., M.S., a twenty-year-old man who worked as a vacuum salesman, was permitted to drive on the grounds that he was supervised by the vocational rehabilitation department and under the care of a physician. Interestingly, the board also reasoned that he needed the car to carry his merchandise from door to door.[128] Similar case-by-case approvals included a man who delivered vehicles for washing at a gas station despite his continued "mild seizures" and a farmer whose license was restored in spite of recent seizures because his son could no longer perform the one-mile drive between the family's farms to provide help.[129]

Women applicants, though they represented a small number of cases before the board, were denied limited licenses more often than men on the grounds of their insufficient need. In keeping with postwar gender roles, physicians explained, "Usually, but not always, a woman finds it easier to accept denial of a driver's license than a man."[130] Yet whereas married women were almost never successful in their applications, unmarried working women more frequently obtained licenses. Board decisions thus reflected gendered norms of economic dependency in marriage more than an idea that women were intrinsically or universally passengers. Among the self-supporting women who gained licenses were factory and office workers employed in areas underserved by public transportation, women who lived in the country but worked in the city, and teachers who taught at schools in outlying areas.[131] Some of these women had experienced seizures as recently as twenty months prior to their hearings.[132] By contrast, S.G., a thirty-year-old woman denied a license in 1954 despite her low incidence of seizures and nearly two years seizure free, was typical of married female applicants. In the board's summary of her case, they stated without elaboration that "there did not seem to be great need for her operation of a vehicle."[133]

Women, in their lesser "need," were also subject to higher standards of control. Immediately following the review of S.G., the married woman denied a license in 1954, for instance, was that of a married man. Like S.G., F.E. had experienced his last seizure less than two years prior to review but was deemed a "proper person for a license."[134] The board documented that he followed his prescription, was under a doctor's care, and, most tellingly, was "reliable as shown by continuous employment."[135] Work and masculinity might therefore rank above other factors or indeed become the very evidence of a person's safety and self-containment. In such cases, one's level of need for self-support—implying, by extension, one's gender and class— became the primary metric of approval. Consequently, as was determined with regard to another married woman's application in 1954, it was often "better to wait to see if [her] seizures can be completely controlled" since there was usually "no apparent need" for such women to drive.[136] For these applicants, even "complete control" might not constitute an acceptable level of driving risk.

Licensing outcomes for African American applicants are less clear. Given that approximately 98 percent of Wisconsin residents were white in the decade following World War II, it is plausible that the vast majority or even all of those who underwent hearings for limited licenses were white.[137] It is also possible that work-related need, correlating most starkly with gender, was the primary axis of approval versus rejection. Unfortunately, while records for Wisconsin's hearings reveal demographics of age, gender, and marital status, they made no mention of race, rendering it a poor case in this respect.[138] Nevertheless, in what remains a fairly closed historical process, race—whether in Wisconsin or in states that subsequently followed its model—would have certainly impacted the subjective criteria and decision-making of medical boards, or perhaps whether hearings were conducted in the first place. Wisconsin's requirement for in-person evaluation, for example, excluded some people living in more remote and rural areas, mostly white Wisconsinites. The requirement, however, may have also discouraged Milwaukee-based individuals, including the majority of the state's Black residents, given an all-white review board that evaluated according to poorly defined criteria of reliability and need. With more married African American women than white women working outside the home in this era, the lack of data also raises questions about whether Black women gained licenses more often or whether African American families were more greatly disadvantaged by the procedure overall.[139]

Ultimately, the aim of Wisconsin's limited licensing law was to determine the capacity of individuals to drive safely rather than to restrict them on the basis of a diagnostic category alone. This, of course, was precisely what reporting laws had done. At the same time, the format of board review introduced calculations that were by nature subjective and inconsistent. The licensing terms that successful applicants met—much like the notions of "control" upon which they were based—were similarly idiosyncratic. Even when granted a temporary license, some people and not others had restrictions placed on the mileage or distance they could travel from their homes. The rationale in these cases was that restrictions would "limit the amount of driving and thus the exposure [of the individual] to the public."[140] Long trips on unfamiliar roads were also prohibited in certain cases since they were said to promote fatigue and stress that could trigger seizures for some people. Overall, the inequitable distribution of restrictions showed that otherwise strong applicants could be deemed more or less capable or susceptible to seizures by the board. Although people with epilepsy were no longer considered unequivocally unsafe, there also remained no firm parameters by which to guarantee their control or safety on the road.

Control, Rights, and Culpability

By the 1950s, drivers with epilepsy were being reclassified as "safe risks"—and increasingly, safe citizens—in various states across the country. Over the course of the decade, limited licensing expanded as a medical and legal practice in more than a third of all states. In 1956, only seven years after Wisconsin first passed its law, twelve states had adopted a review-style protocol to issue licenses to "adjudicated epileptics."[141] By 1960, six more states granted temporary licenses to people with epilepsy on the basis of physician-certified seizure control, while only seven states continued to enforce compulsory reporting laws alone.[142]

The uptake of limited licensing coincided with broader efforts by lawyers, doctors, and lay advocates to liberalize exclusionary laws related to epilepsy throughout the postwar era. By the mid-1950s, for instance, discussions emerged within epilepsy societies pertaining to workmen's compensation, immigration, and marriage and sterilization laws, coalescing in some cases into platforms for state-level reform. In addition to being, as the *Milwaukee Sentinel* boasted, "the first state to give the controlled epileptic a driving license," Wisconsin also held the distinction of being the first to reverse laws that prohibited marriage and committed people with epilepsy to psychiatric institutions in 1953 and 1955, respectively.[143] Credited as more progressive than most states on matters of epilepsy, Wisconsin was particularly well regarded for its innovation of the limited license, which was upheld as the "ideal procedure" for other states to follow.[144] As the *New York Times* reported in 1955, only a year after New York groups began to lobby for a similar law: "Wisconsin has been licensing certain epileptics for years and the experience there has proved they constitute no traffic hazard."[145]

Support for limited licensing laws outside medical communities, including in the press and legal circles, was similarly based on ideas of medical control. In 1956, Roscoe Barrow, legal advisor to the American League Against Epilepsy, and Dr. Howard Fabing, former president of the American Academy of Neurology, published the landmark study *Epilepsy and the Law: A Proposal for Legal Reform in the Light of Medical Progress*. Understanding legal reform as a natural outgrowth of medical innovation, they reasoned that "remarkable progress in treating epileptic seizures calls for a complete reappraisal of our laws and administrative practices governing the issuance of drivers' licenses."[146] Through proper use of anticonvulsant medications, they argued, people with a "history of seizures" could become "reasonable" and even "safe driving risks."[147] Popular endorsement of drivers with epilepsy also drew direct lines between medical progress and reform. Newspapers described the need to "bring discriminatory laws in line with scientific facts about epilepsy," congratulated Wisconsin on its efforts to cast off "age-old superstition," and called for further action to "eliminate legal discrimination against victims of epilepsy as automobile drivers."[148]

As epilepsy entered a new discourse of safety and rights throughout the 1950s, the car and driver's license became potent symbols of the disadvantaged place that those with epilepsy occupied within American society. In 1960, Dr. Lennox, then president of the International League Against Epilepsy, summarized the situation: "A driving license is a means of identification."[149] But it was more than a mere identification card; Lennox also pointed to the ways in which the privilege to drive was enmeshed with one's identity and freedom. An "indispensable element of our way of life," he argued, a car was necessary for work and travel and to "keep the respect of and friendship of peers."[150] As Fabing and Barrow further suggested, denial of this right could be "a serious psychological obstacle to the adjustment of the individual in society."[151] Not driving, moreover, might inadvertently serve as a marker of epilepsy. Unlike other public health restrictions that did not "involve the public's knowing about the condition," those facing driving exclusions would "continually have to give some explanation why he could not have a driver's license."[152] "Denial of the privilege to controlled epileptics," the lawyers reasoned, "creates a serious obstacle to their rehabilitation. . . . Lack of this privilege is noticed in the community and the individual becomes identified as epileptic. Thus, the driver's license law becomes warp and woof of the epileptic's discriminatory sackcloth."[153] Licenses then were not only a symbol of equality—or tool of economic opportunity for young men in particular—but also a means to acceptance, integration, and "rehabilitation."

Still unclear in these recalibrations of safety and inclusion, however, was where responsibility fell for failure: What happened when the seizure-prone driver faltered because seizures occurred? Insofar as support for limited licenses hinged on ideas of medical control, it also presumed *individual* control; one had to be a "controlled epileptic"—a phrase and classification increasingly invoked by epilepsy's advocates—in order to obtain any number of rights or privileges.[154] Yet such a prerequisite also meant that additional responsibilities fell to those with epilepsy. Rather than expectedly subject to periodic losses of personal composure, "controlled epileptics" were implicated and sometimes blamed for seizures, whether because they took their medication incorrectly, did not acknowledge "warning signs," or, in the case of driving specifically, drove after "overburdening" their systems with alcohol, insufficient sleep, or overexcitement.

These issues came into focus when licensed drivers with epilepsy were involved in fatal and nonfatal car accidents throughout the 1950s.[155] Dr. Lennox described the legal and ethical quandary of the "controlled epileptic" who crashed: "Of course, even a controlled epileptic may do serious damage when either in or out of seizure. . . . Unfortunately the epileptic may kill if an attack occurs while driving, . . . What does the law say? Is he responsible and if so did he commit a crime, or was he only negligent, or both?"[156] Although courts across the country delivered several not-guilty verdicts over the course of the decade, criminal responsibility in many cases was assigned to seizure-prone drivers. As in the 1957 New York case of *People v. Freeman*, people were sometimes charged with "negligent homicide" or manslaughter when they ostensibly disregarded the auras preceding their seizures, imbibed even the smallest amounts of alcohol, or simply "knew of the danger involved" in driving a car.[157] Such rulings revealed assumptions about the ability of those with epilepsy to anticipate their seizures and regulate them accordingly.[158] They also suggested higher standards for abstinence and an implicit exchange of full liability for the privilege to drive.[159] Even Lennox ventured that the chances of an accident were greatly diminished if the driver was "aware of his condition and type of seizures, . . . avoids dense traffic, stops when he has an aura, and so on."[160] Given the likelihood of lawsuits, he also recommended that the "epileptic should carry heavy automobile liability insurance against damage to property or person."[161]

At the same time, however, others involved in these debates argued that medicated people with epilepsy were "controlled" to the best of their ability and that any lapses of control resulting in accidents should be pardoned.[162] "Responsible" and "controlled" epileptics, they reasoned, were "unjustly discriminated against"—particularly when considering the socially permissive attitudes that surrounded a rampant problem like drunk driving.[163] Recognition was rising among epilepsy advocates not only that the impairment caused by alcohol was a more extensive and preventable highway safety issue but that alcohol was both widely tolerated and poorly regulated.[164] Wisconsin physician Dr. Roemer frequently described a patient who lived next to a tavern and watched "drunks getting into their cars every night" but who herself was unable to obtain a license until 1949.[165] Others noted that people with epilepsy were often excluded regardless of their driving records and, in fact, prior to ever having an accident. Drunk drivers, on the other hand, were usually apprehended only following an offense. Such logic, to some, seemed equivalent to barring not just those with a history of drunk driving but, further, those recovering alcoholics who had "not touched a drop" in years.[166]

Decades, then, before wider campaigns against drunk driving gained mainstream support, proponents of limited licensing laws contrasted seizure-prone drivers to drunk drivers in efforts to further their cause. Unlike a seizure, use of alcohol, they argued, represented a distasteful "willingness" to lose control.[167] An intoxicated person was "uncontrolled" by choice, whereas the "controlled epileptic" who experienced a seizure was out of control through no fault of their own.[168] Distinguishing between voluntary and involuntary forms of control in this way also rendered a drunk driver "more culpable" for a car crash than a person with epilepsy "since he knowingly brought his incapacity upon himself."[169] Others went so far as to condemn and characterize the "intoxicated motorist who kills in a traffic accident" as a person engaged in an indulgent act of "self-induced alcoholic amnesia."[170] Lamenting the "kid-glove treatment" of drunk drivers, which they saw as a much more serious safety issue, doctors and patients ultimately attempted to parse the slippery issue of control, including the circumstances under which losses of control might be forgivable.[171]

1955: Transparency and the End of Reporting

When Wisconsin successfully repealed its 1947 reporting law in 1955, leaving only the model of the limited license, it did so on the grounds that compulsory reporting increased the problem of concealment. Fear that people with epilepsy would hide their condition, even from their doctors, became a primary rationale for granting licenses in many states, as well as for the gradual abandonment of reporting laws across the country. As a lawyer described to Dr. Crownhart in 1959, "Such laws dissuade many epileptic patients from seeking treatment and encourage them to keep their conditions secret. Concealment in turn tends to increase rather than decrease the number of epileptic licensed drivers on our highways, and thus to decrease rather than increase the likelihood of their cure."[172] Fabing and Barrow similarly argued, "Compulsory reporting statutes drive the epileptic underground . . .

fearing that disclosure of their conditions may result in a deprivation of substantial rights."[173] Likewise, neurologists at a 1956 symposium on accident control in New York recommended ending the practice of reporting so as to avoid "subterfuge and secrecy."[174] Experts thus increasingly endorsed licensing for some people with epilepsy—not because of a trust in medical control or their competent driving—but because it discouraged noncompliant individuals acting in secret.[175]

Notably, the shift away from compulsory reporting also discharged much of the remaining responsibility for control and safety from doctors to private citizens. For reasons of doctor-patient confidentiality and professional liability alone, reporting laws had met resistance from many physicians throughout the 1940s and early 1950s.[176] As reporting laws were stricken from various states' records, however, doctors had no further obligation to report cases of epilepsy unless requested by the patient. Instead, the onus was on individuals to submit their own names to local authorities.[177] Although physicians still regarded themselves as "the most reliable source of information as to who is epileptic," disclosing such information without permission once again became the "willful betrayal of a professional secret."[178] Making the seizure-prone person accountable for his or her own reporting and driving record, they estimated, was beneficial to all, not least the patient, who became "responsible for himself."[179]

How people with epilepsy responded to the adjustments to Wisconsin's laws in 1955 remains less clear. While several were motivated enough to undergo biannual hearings, reactions throughout the immediate postwar period are otherwise poorly documented. Greater organization and advocacy by late 1960s and 1970s, however, does show that some seizure-prone drivers and their families were openly dissatisfied with the contradictions they saw between greater responsibility and restricted rights under continued limited licensing laws.[180] A Wisconsin mother, for instance, who protested the revocation of her son's license in 1979, wrote to her Motor Vehicle Department: "How can you come to our home, which has nothing to do with vehicles, and prove our son had a seizure? You can't, and yet they are the ones who have difficulty getting and keeping a license. . . . This young man reports his seizures. . . . That magic word 'epilepsy' or 'seizure' perks up Motor Vehicles and there goes the license."[181] Echoing earlier critiques of reporting requirements, she also reasoned that whereas drunk drivers had to be proven "guilty" on the spot, epileptics were "guilty before proven so."[182] This, she argued, criminalized and further stigmatized people with epilepsy, limiting them in both their social and vocational opportunities. Highlighting the ways in which her son's good character aligned with review board priorities, she added, "He was an honors student in high school and maintained a better than B average in college. He has never had a traffic violation, an accident, or a seizure while driving."[183]

Frustration with the review board itself also grew during the post-reporting era. By the late 1970s, the Wisconsin Epilepsy Association petitioned the state's Motor Vehicle Department, citing "multiple phone calls" from disgruntled citizens. In addition to inadequate information about the review process itself, numerous people had complained about the board's lack of due process, arbitrary decision-making, and lengthy delays in assigning appointments.[184] Applicants also became more vocal during their hearings, including one environmental engineer who protested a twenty-five-mile radius restriction on his license. When asked how much driving it was necessary for him to do, the engineer queried, "Necessary to satisfy what need?" When asked if he had a "system" for taking his medication, he replied, "My system is to pick them up, put them in my mouth."[185] Whereas the board's goal was to reduce the man's overall driving and ensure that he would drive only in familiar areas, the applicant countered that most accidents happened close to home and that a radius would not actually limit his total amount of driving.[186] Whatever the board did to assess his merit or "reliability," he resisted their scrutiny.

Licensing Risk

By the middle decades of the twentieth century, driving was increasingly at the center of rights-based negotiations regarding seizure-prone people. Before the right to marriage, equal education, or work, driver's licenses and the privilege to drive became early markers of wider, albeit conditional, forms of inclusion. It was also the context in which the idea of the "controlled epileptic" took shape, yet far from the only arena in which claims to control would hold value or opportunity for people with epilepsy in postwar society.[187]

Redefining the safety of seizure-prone drivers represented an ongoing attempt by physicians and lawmakers to impose order and authority both on and off the road. Aiming to mitigate the uncertainties associated with epilepsy, compulsory reporting practices in the 1930s favored identification and blanket refusal of licenses. In this first iteration of safety, physician informants' total removal of "epileptic drivers" from the roads had seemed most secure. Limited licenses then emerged after World War II as a compromise for doctors and driving applicants alike. Medical review boards determined the "reliability" and "need" of individuals—often over seemingly more objective measures, like periods of "certified seizure control." Licensed drivers with epilepsy after 1949 were therefore "safe," but only insofar as they fit narrow definitions of properly monitored and rehabilitated citizens.

Safety, while in name a state of unambiguous embodied control, continued to be about erecting measures that addressed fears concerning the risks posed by unidentified and invisibly afflicted persons. The Kansas City salesman, after all, was an enduringly frightening character: inconspicuous, seemingly innocuous, but potentially deadly. Through the evolution of mid-century laws, people with epilepsy became both institutionally visible and personally responsible for their control on the road. No medical or legal framework eliminated accidents or the commonplace dangers of the car. Yet they did, for a time, make clear who and what was to blame when accidents happened.

III SEIZURES AND CONTROL

The Operable Brain

Seizures, Surgery, and Patient Subjectivity

Jean traced her seizures to a single traumatic experience. At the age of seven, while walking through a field with her brothers, Jean was approached by a man carrying a sack who asked her "How would you like to get into this bag with the snakes?"[1] The girl ran home to her mother mostly unscathed, but in the years that followed, she began to relive the terror of this moment with every fiber of her body. Any inkling of the incident produced screaming, flailing, and the onset of grand mal seizures that left her scared and depleted. Seven years later, in the summer of 1936, fourteen-year-old Jean lay on an operating table at the newly opened Montreal Neurological Institute. A white sheet was attached to her scalp, and a large piece of her skull had been turned back to reveal the right side of her brain to her doctors. She was locally anesthetized but fully lucid. Neither she nor anyone present anticipated that she was about to help find the troubling memory.

The person identified in the medical literature as "J.V.," referred to here as "Jean," is a significant but largely unknown neurosurgical patient.[2] In 1936, Jean was among the earliest recipients of a surgical technique developed by Dr. Wilder Penfield at the Montreal Neurological Institute (MNI) for the treatment of focal epilepsy. During the electrical stimulation phase of her operation, as Penfield "mapped" her right temporal lobe with a low level of electricity, Jean spontaneously recalled the childhood memory and experienced the intense fear that always preceded her seizures. By speaking aloud of this aura from the operating table, Jean not only helped doctors identify the neural tissues that they sought to excise but also offered a window into the mechanisms of the mind. Beyond the motor and sensory responses of twitching limbs or tingling lips that more commonly typified patient feedback during stimulations, Jean's sudden and anomalous vision of past experience, shared with her doctors, invited new questions about the nature and place of human consciousness. The young woman soon captured the career-long interest of her surgeon and an emerging generation of brain scientists. And yet, beyond the sparse details that accompany her initials in medical publications, her story, like those of Penfield's patients more generally, has remained obscure.[3]

Between 1934 and 1960, a total of 1,132 people with intractable epilepsies traveled to the MNI to undergo operations that they hoped would eliminate their seizures.[4] These surgeries represented a new type of treatment for epilepsy and contributed enormously to understanding of the brain. As historians have documented, Penfield and his colleagues gained unprecedented access to the cerebral cortices of conscious subjects through their surgical treatment of epilepsy. In particular, electrical stimulation of locally anesthetized patients yielded some of the first and most detailed brain maps, gleaning much for medical science about the topography and functional anatomy of an otherwise elusive organ.[5] However, while the involvement of Penfield's epilepsy patients in these endeavors is noted, histories of the procedures have focused mainly on the manner in which surgery allowed physicians to extrapolate about the human brain and mind. Less is known about the patients themselves, including the extent of their contributions or—as is relevant here—the personal and broader contexts in which they underwent surgery.[6]

This final chapter explores the role of people with epilepsy as thinking and speaking surgical subjects in new neurosurgeries for epilepsy during the middle decades of the twentieth century. It shows that patients like Jean were vital, albeit unequal, participants in making brain science. Equally, it underscores the importance of centering stories of epilepsy—particularly the therapeutic hopes of patients—in this history.[7] During an era in which wakeful surgeries were rare but increasingly prevalent, physicians and patients constructed the brain through both patients' experiences in surgery and their ongoing postoperative feedback, engaging in an expansive process of discovery that generated previously unobtainable results. As the chapter argues, patient testimony of various kinds constituted essential but highly mediated components of brain research; while patients volunteered their subjective experiences to varying degrees, physicians had to accept and interpret the "data" provided by patients. Penfield, in fact, considered some patients reliable and deemed others whose feedback failed to corroborate medical understandings of the brain untrustworthy. Moreover, despite these tensions and the personal risks patients faced, the chapter finds that Penfield's patients were largely eager to participate in operations.[8] Unlike many nonvoluntary recipients of contemporaneous psychosurgeries for schizophrenia and major depression, most were self-selecting and sought out the surgery for a range of reasons.[9] Not least among their motivations was the pervasive social stigma associated with epilepsy and the perception, shared with their doctors, that seizures were reducible and localizable problems amenable to surgery.[10]

Growing historical interest in the neurological sciences has drawn more recent focus to neurological patients, with several studies examining the ways in which neurological knowledge and patients have been mutually constructed.[11] As scholars within this subfield argue, the centrality of the brain and broader nervous system to modern concepts of the self facilitated unique doctor-patient interactions.[12] Given the underlying importance of sensation and thought in clinical investigations, patient testimony became critical to understanding a range of neurological variations and brain functions. Individual "lesion patients," including famous ones like "Patient H.M.," thus held seemingly tremendous potential to reveal general truths about the brain, leading to their "curious individuality" and even renown beyond the scope of their conditions.[13] Yet, as Stephen Jacyna suggests of what World War I era aphasic or "speechless" patients taught doctors about the brain, the supposed capacity of Penfield's patients to, as he said, "speak secrets" underscores the twofold nature of their role as communicating subjects. For one, Penfield believed that patients' subjective experiences emerged as raw data to be interpreted and contextualized by medical experts. At the same time, much of what was spoken, whether unused or flattened into facts and general representations, constitutes a still largely unexamined human experience.[14] To "speak secrets" thus highlights the asymmetries of the knowledge-making venture and an enduringly elusive patients' perspective on epilepsy and surgery.[15]

To better understand this underexplored social history, including how the brain was construed between doctors and patients, this chapter draws on previously unexamined patient records from the MNI. This research includes a number of Penfield's American and Canadian patients, as well as their stimulation reports, case note dialogues, and letters to and from Penfield from the late 1920s to the 1950s. The chapter also integrates several of Penfield's many scientific publications and informal writings. Although patients' perspectives are unevenly documented and are refracted through institutional records, the sources begin to reveal who Penfield's patients were, why they opted for surgery, and what they gained and lost through these experiences. Collectively, they also show the entanglements of medical research, therapeutics, and the often overlooked dimensions of patients' involvement in science.

Becoming a Surgical Subject

During the late morning hours of Jean's first operation in 1936, Wilder Penfield stimulated points across the surface of her exposed cerebral cortex with a low current of electricity. Jean was awake because her feedback was essential for finding the source of her seizures and for ensuring that brain tissues vital to her functioning would be retained later in the surgery. To locate Jean's epileptic "focus," or the tissues suspected to cause her attacks, Dr. Penfield explored a larger region of her brain. Jean did not feel the electrical impulses upon her impervious cortex, but she did feel the sensations they created in her body. Like the few dozen patients who had lain awake on the operating table before her, Jean felt her muscles spontaneously twitch, flex, and jerk. Describing her sensations through words and gestures as the stimulations progressed, Jean delivered feedback that became part of the written record of her surgery.[16]

After extensive "mapping," Jean made an announcement that caught the attention of her doctors and nurses. "I had a funny feeling," she said, "it feels like an attack."[17] With another pulse at the same location, she stated, "I hear a lot of people shouting at me."[18] Penfield ticketed this area of Jean's brain with a typewritten number. "There they go again," she said when stimulated again, then suddenly wept: "I see something coming at me . . . I feel something dreadful is going to happen . . . Please don't leave me!"[19] Soon after these stimulations, Jean was ready for the next phase of her operation. At the end of a long procedure, a four- by three- by one-centimeter piece of Jean's right temporal lobe, including the site that reproduced the hallucinations and fear associated with her "dreamy state" seizures, was removed and sent to neuropathology. After the successful excision of the identified tissue, Penfield positioned a temporary drain and closed the bone flap in her skull with steel wires.[20]

As a relatively wealthy, white patient from a New York medical family, Jean arrived at the MNI with her parents only a year after it opened in 1934, reporting an acute worsening of her condition. When she first met with Penfield in 1935, her occasional petit and grand mal seizures had developed into more serious episodes in which she screamed, cried, and clung to bystanders as often as four times a day. In addition, she occasionally lost control of her bladder and bowels, had been removed from school following a seizure, and made a scene in church that caused her family deep embarrassment. Her impression that someone was following her was unrelenting, and she neither tolerated nor was improved by available medications, like phenobarbital. For Jean's parents, her community, and Jean herself, the situation created by her epilepsy and seizures seemed increasingly untenable.[21]

Each of Penfield's patients who became a candidate for surgery did so because they had intractable focal epilepsy. Their seizures were understood to originate from a single region of the brain and were inadequately controlled by other means. Penfield and several of his contemporaries believed that such foci were generally due to congenital abnormalities or acquired injuries.[22] Instrumental and difficult births, childhood illnesses, fevers, falls, and comas were all commonly cited sources of cerebral cicatrix, the brain scars said to cause recurring seizures. Many of Penfield's patients also had histories of traumatic brain injury, including collisions with cars and sporting equipment, falls on stairs and from balconies, and even gunshot wounds.[23] Others had no such identifiable etiologies, but the clinical presentation of their seizures—whether they began with a predictable turning of the head or curling of a hand—suggested a localizable focus in the brain. The presumed concentration of such tissues appeared to make seizure cessation possible through surgical removal of the cicatrix or region in question.

A relatively diverse and diffuse group, Penfield's patients were referred to him by general practitioners and specialists all over Canada, the United States, and occasionally Europe and Central and South America. Most were therefore displaced from their homes and local institutions when they settled in Montreal for treatment.[24] In addition to the shared experience of unrelenting seizures, many were in the first three or four decades of life, though older adults and several children were treated. Most were white, though from a range of class and ethnic backgrounds, and men and women appear to have been operated upon equally. During its early years, some patients arrived at the MNI through Penfield's former professional ties at New York's Presbyte-

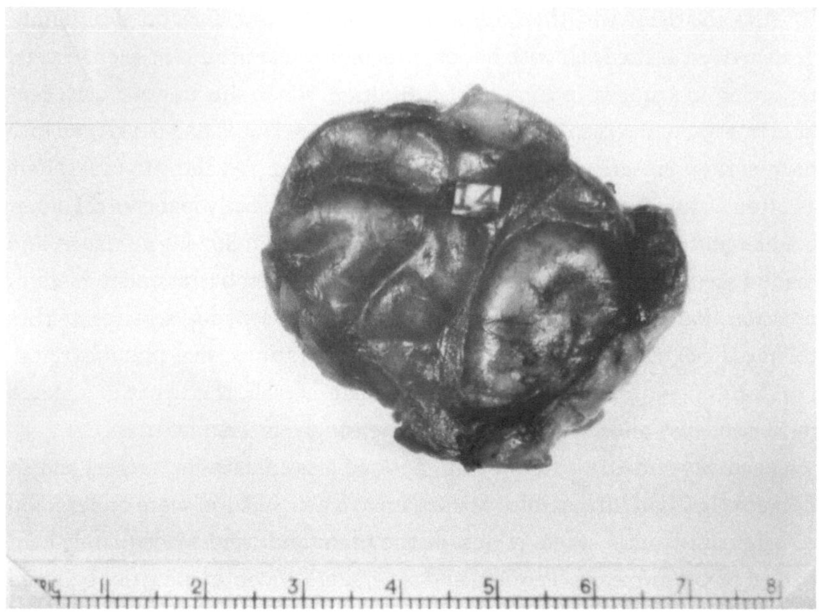

Neuropathology report showing removed piece of brain and remaining paper ticket from stimulation, 1930s. Image courtesy of the Osler Library of the History of Medicine, McGill University

rian Hospital and his private practice on the Upper West Side.[25] Others were the friends and patients of eminent colleagues, like Drs. Stanley Cobb and William Lennox at Harvard.[26] Upper- and middle-class patients could take rooms in the Institute's private and semiprivate wards at a cost of $300 per operation and stay.[27] The greater number who entered as public patients used the MNI's public beds, subsidized by the province and city, for the lesser fee of $10 total.[28] Although wealthy patients were permitted in the public wards, "the only drawback to this," as Penfield explained to the parents of one young patient, "would be that he might see other epileptic seizures."[29]

At the same time, few of those who consulted with Penfield became candidates for surgery. Even in identifiable instances of focal epilepsy, surgery was attempted only when cases were extreme and all other avenues of treatment had been exhausted. Those meeting with Penfield, for instance, had usually tried and failed to control their seizures with Luminal and phenobarbital (as well as Dilantin after 1938), patent cures, and special diets. Typical was a young man from New York with a history of head injuries who came to the MNI in 1937 after a decade of pursuing various dietary and glandular therapies prescribed by the country's leading physicians.[30] Such histories did not guarantee surgery; in several similar instances Penfield advised patients to "postpone surgery until seizures become more severe."[31] In some cases when patients had been approved for surgery, their personal physicians tried to contest the relatively new procedure. In 1931, for example, Dr. Henry Rawle Geyelin, founder of the ketogenic diet for epilepsy, sent a harried telegram to Penfield just days before Penfield was scheduled to operate on his patient, stating: "Do not believe she will prove suitable case for you."[32] Penfield responded only afterward with a defense of his therapy: "We operated on [Patient V.G.] today and there was a definite focal area in the frontal temporal lobe, stimulation of which produced perfectly a convulsion. . . . I have done a local excision of the focus . . . [and] the patient is in very good shape."[33]

With a mortality rate nearing 5 percent in the 1930s, Penfield's operations were complicated and risky procedures that could prove challenging for patients.[34] Beginning in the early morning hours, surgeries lasted approximately eight hours without a break for surgeon or patient. After having their heads shaved by the Institute's barber, patients were scrubbed and wheeled into the operating room, administered local anesthetic, and lay awake while a drill and saw made a trapdoor shaped "flap" in the designated part of their skull. Patients remained alert while doctors stimulated and mapped the exposed region of their brain, ticketed critical sites with small pieces of paper, and identified the neural tissues causing their seizures. During and after surgery, patients exhibited a variety of responses. Jean, who was said to be "prone to hysterics," was in fact "fairly game through the operation."[35] Others, in retrospect at least, expressed nonchalance, such as one patient who reported twenty years after his surgery that compared with his more recent kidney problems, "an operation on the brain was nothing."[36] Many, however, were openly fearful, including one patient who pleaded during the procedure "Oh God, please watch over us" and later admitted "I was so scared."[37]

Following surgery, most patients stayed at the MNI or the adjoining Royal Victoria Hospital for three to four weeks, though less is documented of this period of convalescence. Only one man, a private patient who was unable to speak for several days following the removal of a cerebral scar in 1933, kept a known journal of his postoperative experience. At first, the man

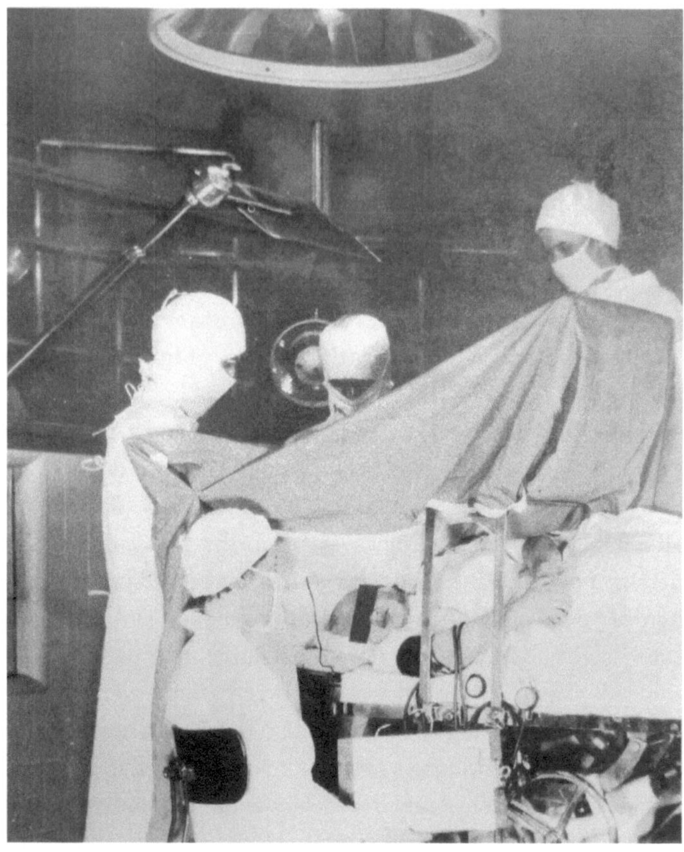

Anesthetized patient surrounded by the surgical team at the Montreal Neurological Institute, ca. 1940. *New World Illustrated*, May 1940

noted only his consumption of ginger ale, the arrival of "fifteen American roses," and hearing Rudy Vallée on the radio. As the temporarily aphasic patient's strength grew over days and weeks, he further detailed "getting my correspondence cleaned up," his need for his secretary, and his appearance before a doctors' convention. "Penfield described my operation," he summarized, "used lantern slides, and asked me questions."[38] In his final days at the MNI, he telegraphed his mother a request for "about three to four hundred dollars" to pay his medical bills.[39] For those reaching the end of a long hospital stay away from work and family, the desire to return to life as usual rather than the surgery itself could be the main focus.

Speaking the Brain's Secrets

The foremost aim of performing epilepsy surgeries, for both doctors and patients, was to limit or eliminate seizures. Yet such procedures also revealed insights into the brain at a time when it was receiving significant attention as a promising subject of research. As Harvard neuropsychiatrist Stanley Cobb wrote to Penfield just prior to the opening of the MNI in 1934, "Surely the brain is the most intricate and most wonderful of all the creations of God. Understanding how the brain functions, to the end that misfunction can be corrected, is the most important and the most neglected of research today."[40]

Prior to new research and neurosurgical developments in the late interwar period, surgeons had not been able to interact directly with the brains of patients to the same degree. Better management of intracranial pressure, for instance, had more recently allowed them to engage in longer and more involved operations. Without the aid of visualizing technologies, knowledge prior to this time had been derived primarily from the study of human cadavers, animals, surgery itself, and, wherever possible, through clinical observation of the cognitive and behavioral changes in various neurological patients.[41] Most notable were those patients who sustained severe cranial injuries, including gunshot wounds and compression fractures, as well as famous patients like Phineas Gage, who lived for twelve years after an iron rod passed through his skull in 1848.[42] Even in 1928, when Penfield operated on his own sister, Ruth, and discovered a tumor in her frontal lobe, his observations of her postoperative functioning during her remaining years of life yielded information about that particular cortical region.[43] Stimulations producing rudimentary mappings of the human cortex were largely novel when Penfield used the technique. The first had occurred on a smaller and more experimental scale among the epilepsy patients of German neurosurgeons Fedor Krause and Otfrid Foerster just before and after World War I.[44] The brain, in other words, though still a relatively mysterious organ, appeared increasingly knowable and traversable to mid-century investigators.

Insofar as epilepsy was characterized by the recurring symptom of seizures, it seemed to offer special occasion to learn about the brain and its functional anatomy. As physicians since the mid–nineteenth century had recognized, seizures provided them with something of a "natural experiment."[45] "Epilepsy before all morbid processes," Penfield summarized, "provides us with multiform experiments upon the brain of man, for epileptic discharges follow out and delineate its functional systems from the simplest to the most highly integrated."[46] Without observing the brain directly, the "abnormal" discharge of electricity during a seizure, when read through its effects upon the body or behavior, had revealed to neurologists like John Hughlings Jackson as early as the 1860s where certain structures and functions lay.[47] Understanding, moreover, that the electricity discharged by an electric current replicated that discharged during a seizure, Penfield's pre-excision mappings of the cortex allowed him even greater command of these "experiments." Focal seizures, believed to originate in specific but different locations of the brain for each patient, needed to be stimulated to be found. Operations to cure focal epilepsy patients of their seizures therefore presented seemingly endless information about how the brain worked.

Yet when Jean underwent her first operation in 1936, epilepsy had only recently been defined as a neurosurgical problem in otherwise healthy and uninjured people. Brain surgery, a conservative arena, was in general reserved for the emergency treatment of life-threatening traumas and tumors.[48] For epilepsy, doctors into the early twentieth century typically attempted operations only in cases of suspected tumors or decompression wounds, where removal of tissues or bone fragments causing seizures proved necessary. Otherwise, the risks associated with operating were too great. By the early 1920s, however, technical developments and dropping mortality rates created new possibilities for surgeons and patients.[49] Penfield, then a neurology resident at Presbyterian Hospital in New York, was part of this changing culture of brain surgery, making experimental wounds in laboratory animals in order to study the healing process of cerebral tissues. He subsequently traveled to Breslau to study with Dr. Foerster, who was at the time operating on German soldiers with seizure-inducing skull fractures.[50] By 1927, Penfield returned to New York, where he operated for the first time on a traumatic epilepsy patient to remove an area of scarred brain.[51] It was not until after he emigrated to Quebec in 1928, however, and with the opening of the Rockefeller-funded MNI in 1934, that he pursued epilepsy surgery nearly full time.

Although reading brain function through seizures, as well as through other "abnormalities" or lesions, was an established feature of modern neurology, detailed electrical stimulation of the cerebral cortex as a preoperative measure—a technique Penfield learned and further developed from Foerster—was not yet common. As he explained to the American Psychiatric Association in 1949, the practice was an opportunity born of therapeutic necessity:

> In the treatment of certain conditions, especially focal epilepsy, it becomes necessary for the neurosurgeon to expose a patient's cerebral cortex under local anesthesia. During such procedures the subject comes to lie quietly on the operating table, fully conscious while the surgeon proceeds with his task. It may be advisable to stimulate the cortex with a gentle electrical current. Such stimulation is able to activate certain areas of the cortex. . . . [Patients] feel a tingling in the hand or foot, . . . see gross lights, hear simple sounds, or smell crude odors.[52]

This exploratory function of stimulation served patient outcomes and, within only a few years, also contributed to detailed maps of the brain. In 1937, Penfield gained recognition for diagrams that spatially linked sections of the brain to localized functions within the body and for his homunculus, a human-like figure in which each part of the body was scaled according to its cortical surface area in the brain—a perspective that yielded, for instance, oversized hands and lips.[53] These models significantly expanded earlier maps, including those of Victor Horsely and Fedor Krause from 1909 and 1912, respectively. A fact less well recognized, or at least foregrounded, about Penfield's schemas was that they represented a composite of the electrical stimulations taken from surgical epilepsy patients.

While Penfield became famous for his functional representations of the brain, his most significant "discovery," and indeed lasting fascination, was what he understood to be instances of "stimulated consciousness" among a small subset of his temporal lobe epilepsy patients. As Alison Winter explains in her study of memory, Penfield produced unexpected hallucinatory and auditory effects in a number of his patients over the course of his career. The first was a woman who, during the stimulation phase of her operation in 1931, witnessed herself giving birth to a baby girl more than twenty years prior.[54] In subsequent years, other patients just as spontaneously had visions of themselves in school bathrooms or on street corners in other cities, heard familiar voices calling to them, or, as in the most cited case of all, believed a radio was suddenly switched on in the operating theater.[55] These strange instances, which occurred only to aid the "scientific understanding and treatment of epilepsy," moved Penfield's findings into the realm of psychiatry and the mind, while simultaneously evoking for him deeper associations with spirituality and phenomena that could not be explained by science alone.[56]

Among these rare cases of "stimulated consciousness," Jean was Penfield's first patient to show that an electrical impulse could activate what he called a "record of past experience." Immediately following her operation in 1936, he noted in her file with puzzlement:

> I cannot explain this, but without any warning she gave me the same response. . . . The very curious response of hearing a large number of people shouting could be obtained in a more or less straight line from the first temporal convolution backward. . . . This sound of voices might well be a continuation of the dream which she has of herself walking in a field. . . . It is possible that this line is the highway over which the epileptogenic discharge passes forward ordinarily.[57]

Although Jean's "continuation of the dream" was unusual, Penfield's sense of its meaning was consistent with his understanding of the brain as a place with identifiable locations and even "highways." Just as he thought that electrical stimulation could replicate an epileptic discharge, he interpreted this event to mean that it could also reproduce the "neuronal action" of a memory.[58]

Though few and sporadic, cases of "past experience" such as Jean's invited questions among a growing number of experts about the higher orders of consciousness and, seemingly, about the very core of humanness. To these unknowns, Penfield and his colleagues applied the same localization sensibilities that they had to problems of reflex and sensation, compelled less toward understanding the metaphysical nature of the mind than to finding its physical basis in the brain.[59] In a 1952 article, Penfield quoted Stanley Cobb musing, "And where is the place of understanding . . . Where is the 'locus of the mind?' "[60] In 1960, he persisted: "Why do these experiential responses only occur in epileptic patients. . . . Why do they only occur in temporal lobes?"[61] Like particularly illuminating instances of aphasia or amnesia shared by doctors, Penfield's stimulated-consciousness patients created much excitement within the medical and scientific communities.[62] Given their far-reaching implications and applications, his results spurred interest among psychiatrists, psychologists, neurologists, neurophysiologists, and other neurosurgeons, who amounted to a growing array of brain and neuro-adjacent experts by the middle decades of the twentieth century.

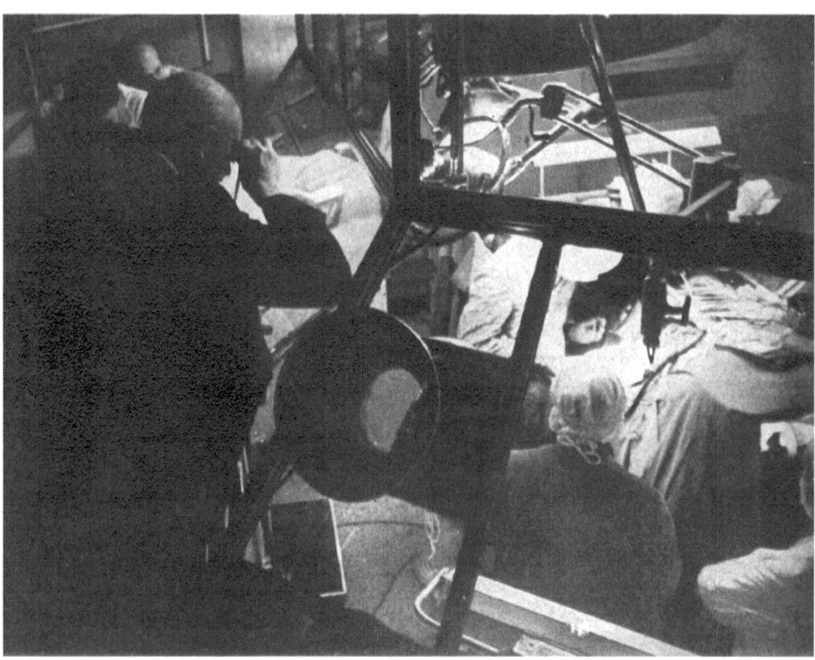

Wilder Penfield overseeing a surgery with opera glasses from the Montreal Neurological Institute's operating theater gallery, 1956. Note the speaking tube through which he can communicate with the surgical team during procedures. Courtesy of *Maclean's* magazine

New technologies and spaces also fueled Penfield's findings. The MNI's state-of-the-art surgical theater, which he helped design in the late 1920s, came to integrate the most up-to-date features of modern surgery and pioneered several others. A photography chamber beneath the spectator gallery used mirrors and magnifying techniques to unobtrusively photograph each procedure. Dictation and recording devices in the operating room, as well as ever-improving brain ticketing and documenting practices, further facilitated patient study.[63] By the late 1930s, EEG was also incorporated into surgery through the work of Herbert Jasper.[64] Few details thus escaped Penfield's attention. In some cases, patients and their transcripts became "scientifically pertinent" only months or even years after their operations. As Penfield acknowledged of the first "stimulated auditory response" patient, for instance, "we only came to realize several years later that it was the auditory area."[65] Consequently, when Jean bore witness to her inexplicable vision, it could be appreciated in the moment, but it was also captured for infinite revisiting.

By the 1950s, Penfield's work involving the temporal lobes gained the attention of popular audiences. Magazine and newspaper articles, however, often deemphasized the therapeutic purposes of his surgeries, focusing instead on the gains that the lone surgeon had made toward localizing brain function and uncovering its mysteries. As one 1958 article began, "A central problem of brain research is to discover where and how we store remembrance of things past. . . . One of the most significant of these investigations is taking place at the Montreal Neurological Institute, where surgical studies on localization of function in the cerebral cortex have been conducted for the past 25 years."[66] Reflecting on these "studies," it continued:

> Taking advantage of the fact that brain surgery is possible with use of only a local anesthetic, Dr. Penfield's group have electrically stimulated the exposed cortex in more than a thousand patients operated on for neurological disorders. From the fully conscious patient, they have amassed a series of provocative observations. Today's knowledge of the function of the human cortex has been gained largely from such operations.[67]

This article, like others both before and after, did not name epilepsy—or the participation of the unspecified and undifferentiated "fully conscious patient." Clinical motivations and aims were thus also obscure. The scientifically extractable knowledge of "brain research," on the other hand, remained front and center.

Operating Room Dialectics

Despite the necessary act of surrender to their surgeon, Penfield's patients were actively engaged in the process of surgery and brain exploration. Their responses were critical both within and outside the operating room, and Penfield acknowledged the contributions of what he called the "feeling patient." In a 1947 lecture to an audience of medical students and faculty, Penfield described the surgeon-patient relationship as one of collaboration:

> It is important to realize what may be going through the mind of the patient as he lies on the operating table. He has no means of knowing when the stimulation is being carried out unless he is told. . . . Although he [the surgeon] has the brain open before him and commands the means and paraphernalia of science,

and though he can make the patient feel and see and hear and move, yet he cannot make him think, he cannot make him believe or decide anything. The man
who lies upon the operating table beneath the sheet also listens and wonders
and tries to understand. Between these two lie the secrets of the function of the
brain.[68]

Although Penfield in some ways counted the brain among science's "para-
phernalia" and provided his own surgical perspective, he also conceded that
the brain was no ordinary object or body part and that the patient could not
simply be manipulated. Rather, complex "secrets of the brain" emerged between doctor and patient. Just as Jean's narrative about the man with the
snakes represented some inner recess of her mind—inaccessible without her
sharing—Penfield recognized that there were dimensions of patients' experiences that were scarcely comprehensible without their cooperation and insight. Simultaneously, however, it was the model of a "feeling" patient paired
with the surgeon's "command" that described these events. Notwithstanding
Penfield's acknowledgment of patients' roles, his characterization of them as
"feeling" rather than "knowing" revealed the hierarchical orientation of the
relationship; while patients could "speak secrets," the physician had to direct the brain to produce its effects and then had to be the one to understand
them.

Given patients' critical role in epilepsy surgeries, it was essential that
they be "reliable" in describing what they thought and felt.[69] Penfield deemed
patients in aggregate sound judges of their own stimulated experiences, yet
individual patients could be considered more or less trustworthy. Not unlike
those who would apply for and gain driver's licenses, those whose experiences aligned with physicians' expectations were deemed intelligible and
thus "intelligent" sources of information. Jean, like many patients who had
a certain status and whose responses corroborated current medical knowledge, was a trusted respondent and collaborator. After her 1936 operation,
Penfield reported that "she is quick, alert, and intelligent. Her responses are
absolutely reliable."[70] Others he regarded with suspicion. Another patient he
operated on that year, for example, was described as "quite normal" and
fairly cooperative but ultimately "not a particularly good witness" because
he "sometimes reported sensations here and there which had no relation to
anything."[71] In such instances, the patient's subjective testimony was labeled
"inaccurate," and no "secrets" were revealed.

Postoperative feedback was equally essential in constructing knowledge about the brain. As with patients' responses during stimulations of the cortex, letters and questionnaires solicited after surgery generated usable data for Penfield and researchers at the MNI. Staff casually encouraged patients to send "just a Christmas card with a note that you have had no trouble of any kind," while also stating that such contributions "make our record here of max value in studying these problems."[72] Particularly among Penfield's patients who produced a "record of past experience" during surgical stimulation, personal accounts and interpretations following surgery could be highly informative. At the same time, it was up to physicians to decide whether such submissions were "relevant" or not. Like formal records from the operating room, letters and questionnaires presented another instance in which Penfield and his colleagues determined which sensations or seizures were anecdotal and which were of "max value" in understanding epilepsy and the brain.

In correspondence, however, some patients appear to have recognized and moved between categories of patient and scientific informant, if not scientific authority.[73] Two years after an operation on her right temporal lobe, one patient wrote to Penfield of her continuing seizures and visual auras of "past experience," like those she experienced before each seizure and during her stimulation. Adopting the mode of investigator, she referred to herself in third person, as "the patient": "The attacks are much less severe. . . . The patient is unconscious for about forty seconds. The eyes do not roll but rather the lids are closed and blink rapidly. There is no froth at the mouth or any twitching. . . . The aura is ever difficult to explain. The attacks occur upon the <u>recollection</u> [underlining in original] of consecutive dates or figures. . . . The patient is always in the same locality mentally when the attacks occur."[74] Closing her letter on a personal note, she returned to the first person, modestly expressing the hope that her input was medically pertinent: "I hope I have been of some small assistance in the marvelous work which you are doing. Please allow me to help further if I can."[75] Her letter at once conveyed a personal narrative of illness, interpreted symptoms and seizures in a clinical manner, and underplayed her status as a key participant in constructing the brain.

Other "record-of-past-experience" patients became aware of the importance of their cases and contributions through media sources. The most frequently referenced among them was the woman who believed she heard a radio playing during the stimulation phase of her 1949 surgery.[76] Two months after her procedure, she wrote Penfield a letter regarding a popular news story on an operation she soon recognized as her own. Her letter matter-of-factly stated, "I am enclosing a clipping from a local paper. Now I know why you asked me about that song almost every time you saw me."[77] The article, "Bared Under Anesthesia: Nerves Store Memory," which reported on findings Penfield had presented to the American Psychiatry Association, focused not on the patient but on the nature of human memory: "The brain carries memories in a complete double filing system," it explained. "There is one on each side of the head. These are called temporal lobes. A stenographer whose brain was bared under local anesthesia so that she remained conscious exclaimed when one part of it was touched that she could hear music. . . . She did not know until after that the music came from her own head."[78] Despite the article's minimization of the patient's understanding of the event, Penfield had been eager for more of her perspective. In a follow-up letter, he inquired, "Is your memory for songs just as good now as it ever was? Have you ever thought in the past that music would precipitate one of your attacks? If you sit down now and imagine that song, is it just as real to you . . . as it was in the operating room where I recalled it to you? Please think this over and write me your opinion about it."[79] Patient self-interpretation and reporting were thus crucial features in building knowledge about the workings of the brain. While decoding its mysteries remained the physician's role, new understandings were often derived from an ongoing exchange between doctor and patient and not just from a single, expertly manipulated stimulation that occurred in the operating room, as was so frequently suggested.

Although it was Penfield who ultimately defined the questions and derived answers from his interactions with patients, patients did not always acquiesce to such unidirectional models of research or doctor-patient relations.[80] When sharing insight about her case at the National Academy of Sciences in New York in 1957, three years after her second operation, Jean highlighted the gaps that could arise between medical perspective and patient testimony.[81] Before Penfield gave his lecture, Jean recounted to the room of physicians and scientists her memories of seizures and experiences during both operations. Though deferential in her participation and delivery, the details that emerged from this personal narrative were slightly different, more extensive, and in some respects less "dreamlike" than those in Penfield's many notes and published accounts of her case. Rather than a "meadow" or "field," Jean now starkly recalled the edge of a weedy lot in Brooklyn. The man emerged not from the grass, but from an adjacent shop, and he was much older, smaller, and more shabbily dressed than indicated in the published story. The sack said to hold the menacing serpents was not a sack at all, but a regular shopping bag.[82] Jean's account also imparted personal details not reported in the medical literature but nevertheless significant to her experience. She described, for instance, the disruption that her seizures caused leading up to her operations, being forced to leave her friends at school, and the challenges that the potential for seizures posed in daily life. Patients could therefore speak "legitimate" secrets of the brain—in keeping with Penfield's "neuronal records" and "filing system" framework of memory—while also emphasizing more individually resonant narratives of their illness and disability.

Seizures and Surgery

More than convenient docents to the brain at a moment of surgical possibility, Penfield's patients were first and foremost there to be treated, and even "cured," of something that society labeled both a disease and defect. Jean, like others who underwent operations, was attempting to rid herself of seizures that were frequent, severe, and poorly addressed by medication. Patient motivations, however, ran deeper than the desire to end seizures or to manage the potentially fatal health risks associated with intractable epilepsy. As voluntary patients, people came to Penfield to "take control" of their lives in climates both fearful and intolerant of their conditions. For instance, only a decade prior to Jean's surgery, two Canadian provinces and several American states adopted eugenic sterilization laws that expressly included epilepsy, while many more passed legislation prohibiting marriage. Across the United States and Canada, people with epilepsy were also institutionalized, were turned away from schools, and had difficulty obtaining and maintaining employment.[83] Although new research in the 1930s and 1940s began to extricate epilepsy from theories of heredity, the notion of genetic taint persisted, as did the shame and secrecy associated with it. In short, the desire to conform to a society that abided neither epilepsy nor seizures created a context that could make patients eager, even desperate, to operate.

Even before Penfield began performing epilepsy surgeries, in fact—and particularly before the widespread uptake of anticonvulsant medications—there was evidence of patient and family demand for surgical interventions. In 1928, for instance, when Penfield briefly left his position at Presbyterian Hospital to study with Dr. Foerster in Germany, he was followed by a wealthy New Yorker who insisted he operate on her sixteen-year-old son. Penfield declined, but her alleged persistence combined with a sizable endowment to the Institute culminated in her son's becoming one of his first surgical epilepsy patients in 1930.[84] Soon after the operation, as Penfield would later recount in his autobiography, the young man moved into his own apartment, attended Columbia University, and, in an act of defiance against his parents, dropped out to get married. Jean's family similarly advocated for her surgeries. Before her second operation in 1954, her doctor wrote to Penfield: "It would be a good idea to completely reveal the problem in the light of new knowledge. [Jean's father] is very anxious to have you do this at your earliest convenience."[85] Less privileged patients and their families also pursued procedures and exercised their medical connections to the best of their abilities. In 1931, a family physician wrote to Penfield of a woman who was "rather anxious to have her sister operated upon by you. . . . It is through the insistence of her sister that I am bothering you with this letter."[86] For many patients and their families, the apparent need to operate seemed to outweigh the risks or fears associated with surgery, indicating the gravity of epilepsy as a medical condition and social category.

Indeed, in addition to concerns about the long-term impact of uncontrolled seizures, interest in Penfield's surgeries reflected the consequences one might face in periodically and unpredictably losing bodily control. Jean's mother, for instance, came to the MNI in 1936 expressing dismay about the highly emotional and embodied presentation of her daughter's seizures. Her "hysteric" outbursts in public and tendency to cling to bystanders were especially alarming. As Penfield wrote in 1939, "From a physiological point of view epilepsy may be defined as tendency to periodic involuntary neural explosions. For the patient on the other hand it is a state of continuing dread interrupted by recurrent attacks of involuntary behavior."[87] Like the "involuntary behavior" that both Penfield and Jean's mother understood as "physiological," the "dread" Penfield referenced as the patients' burden to bear was similarly presented as an unavoidable fact of the disease. Dread, in other words, was overlooked as a social and thus potentially alterable component of the experience; instead, it made sense to intervene directly upon the brain to address both seizures and the apprehension they elicited.

Many of Penfield's patients seemed, in fact, to accept and even embrace surgery as an avenue for treatment and social improvement. In addition to suggesting a less stigmatizing, somatic basis for a condition still strongly associated with mental illness and low intelligence, participation in such surgeries also signaled that patients were courageously taking charge of their illnesses. Spelling out the possible gains of operations in personal and material terms, one Pennsylvania patient who eventually underwent a left temporal lobectomy wrote,

> I have continued to take the seizures with all the heartache they entail, trying to hide my condition for fear of losing my work, desertion of my wife because of my condition and now, I find they not only leave me with a severe headache and upset stomach, but my entire body is so sore I can barely move for several days— this is jeopardizing my livelihood. . . . Do you think you can help me become a normal person?[88]

Others, like an office manager in New Brunswick, Canada, expressed the desire for "normalcy" differently but were equally committed to operating. As the man conveyed to Penfield prior to his surgery in 1942, he did not mind that Penfield "could not guarantee success."[89] "Although I suffer no pain," he wrote, "nor in any physical way am handicapped by this trouble, there is a certain mental drag that I am afraid is changing me from a man to a mouse."[90] Still others, despairing and fatalistic, understood their surgeries in starker terms, stating "I wish it would either kill or cure."[91] Although such admissions exposed a range of outlooks about surgery, implicit in all was the assumption that to have seizures was an unacceptable way to live or be.

Requests for Penfield's procedures throughout the middle of the twentieth century also attested to the growing vision of brain surgery as the answer to a wide range of medical and personal problems.[92] Post–World War II media coverage in particular depicted surgery as a miraculous panacea and the medicine of the future.[93] For instance, a 1948 article in *Time* magazine told the story of a young woman who underwent an eight-hour brain operation at the MNI that "cured" her of the seizures that she purportedly developed as an infant after falling from her high chair. The account emphasized the ease of the operation, the "saucy cone of green felt" that the patient wore

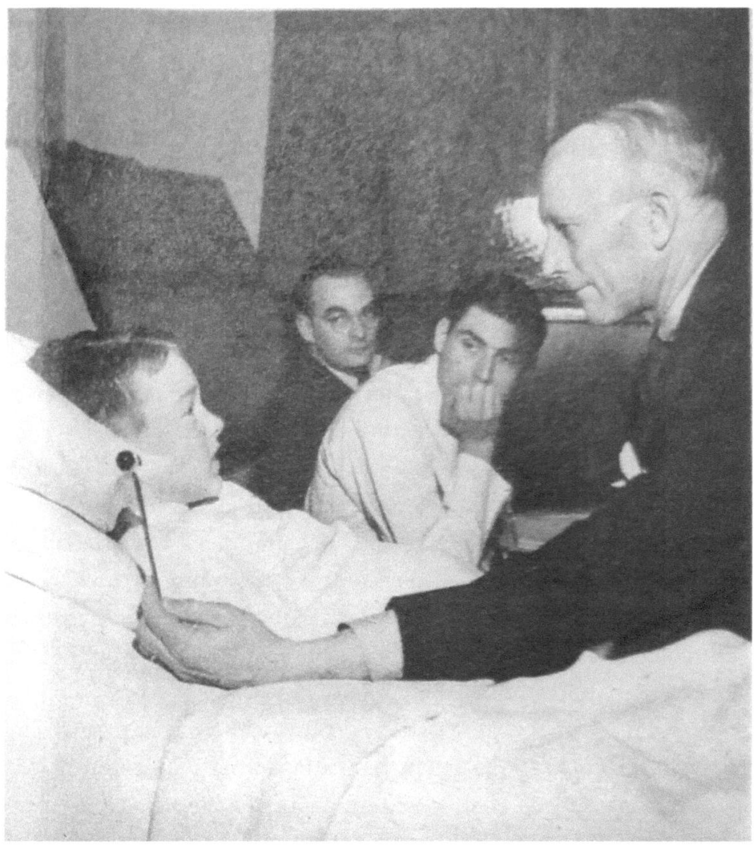

Wilder Penfield examines a young patient during ward rounds at the Montreal Neurological Institute. *Montreal Standard*, Jan. 11, 1947

postoperatively, and Penfield's ability to treat "apparently hopeless cases referred by doctors everywhere."[94] Downplaying the seriousness and selectiveness of surgery, "A Chance for Elizabeth" also presented an unrealistically "routine" procedure and fast recovery for patients. Similar to parallel coverage praising the lobotomy, an eventually abandoned procedure for which Egas Moniz won the Nobel Prize in 1949, the article highlighted the speed and efficiency of the procedure. Such representations underscored the excitement surrounding new approaches to the brain and spoke to a variety of readers.

When articles like "A Chance for Elizabeth" were published, Penfield was inundated with letters from people who wished to obtain his services—

either for themselves or for loved ones. Since articles were generally euphemistic about epilepsy, or failed to mention it altogether, those who wrote described an array of poorly addressed physical and mental health issues. People with epilepsy, however, as well as their parents, siblings, spouses, employers, and solicitous members of the community, proved to be attuned readers, contacting Penfield with the greatest frequency. A Washington, DC, man, for instance, writing "out of sheer desperation," explained, "I have a brother with epilepsy. . . . His attacks on some days are very frequent. . . . As a result the public schools have refused him, and no one else can take time with him. . . . So, the State has put him in a penal institution because he is incurable 'sick.' . . . Needless to say, my mother and the rest of the family are heartsick."[95] Others pleaded with Penfield or described "being robbed of all chance of a normal existence," stating that either themselves or their kin went unimproved by medication.[96] Showing that attitudes surrounding epilepsy and fears of public seizures had not improved, still others sought something "better" than mere "seizure control"; they wanted a cure. The MNI, which then had a waiting list of more than 120 patients and strict criteria for surgery, could not typically accommodate such inquiries and replied by telling letter writers to consult with their local physicians.

Those who became surgical patients of Penfield often expressed similar desperation and frustration with their circumstances, both before and after surgery. Patient records reveal men and women who had seizures in the workplace and lost their jobs. Children were often expelled from school after a convulsion in the classroom, on the playground, or in the presence of another child. In letters, parents of patients described being "nearly worried to death" or "at wit's end" over their children with epilepsy.[97] Many more enumerated the various social costs associated with their illnesses. One patient had "fallen in love for the first time and want[ed] to get married."[98] For others, it was the desire to go to college or join the army.[99] Penfield was generally attuned to these struggles and, like his patients, imagined surgery for eligible candidates as the best and most direct course of action. Insofar as epilepsy remained an individual and operable problem, it was a path that doctors and patients ventured to take.

Outcomes and Intimacies

To undergo surgery, most patients had to be far away from their immediate social networks, endure prolonged hospital stays, and, for reasons of self-preservation, replace whatever apprehension they had about surgery with trust in their surgeon. The ongoing need for periodic follow-up, whether operations were "successful" or not, could also extend doctor-patient inter-actions—often over many years or even decades. Several patients wrote Penfield letters before and after surgery, and Penfield maintained much of this correspondence. Letter writers were self-selecting, wanting perhaps to appease their surgeon or establish lines of communication should they need further treatment or care. Such letters reveal dimensions of the procedure and of these relationships that rarely appear in accounts of Penfield's quest for knowledge about the brain.

Outcomes of epilepsy surgeries performed at the MNI were variable, yet many patients who wrote to Penfield after their operations expressed grati-tude and feelings of personal connection with their doctor. One patient, for instance, writing from a ranch in Arizona where he convalesced in 1936, said, "I seem to find words inadequate enough to justifiably express . . . how grate-ful I am for the good you have brought me."[100] A twelve-year-old boy from New York, a year after his operation, wrote, "I enjoyed the wonderful pres-ent you gave me and it was one that I needed."[101] The boy, who eventually experienced a recurrence of his seizures, explained that he was doing well in his schoolwork, had had "a dozen haircuts" since his surgery, and even joked that Penfield should "look through [his] medicines carefully to be sure that [he] did not use hair tonic, for that too is a dressing."[102] Though by no means unanimous, such responses were common—even when results were less than auspicious.

Penfield, in turn, was warm and appreciative in his letters. In his replies, he often wrote "I am always glad to hear from you" or "I have never known anyone with more comeback than you have."[103] When one of his patients, a young farmer from Vermont, died from infection two weeks after his oper-ation, Penfield was deeply disturbed by the loss. As he explained to a col-league in Montreal, "I became very attached to [patient] and his death is a great blow to all of us."[104] In a letter to the deceased patient's mother, he similarly explained, "He . . . appealed to me at once and I am sure he would have been a very fine man. . . . It was a great joy to me to talk to him."[105] The patient's mother replied, "[He] took a great liking to you Dr. Penfield, and told me you were coming to work on his farm. He felt very much honored to have you say that. . . . We understood he was very brave during the operation and cooperated in every way with the Doctors and nurses."[106] In addition to communicating clinical aims and the costs of surgeries, such exchanges show a side of neurosurgery often absent in accounts of Penfield's work.

Beyond corresponding, Penfield sometimes intervened in patients' personal matters, writing on their behalf to immigration officials, employers, and insurance companies. Vouching on several occasions for the competency, safety, and health of his patients, he might explain that someone was "quite capable of carrying out office work efficiently" or that "occasional minor attacks of dizziness would not serve as any handicap in this type of work."[107] For those who encountered problems crossing the US–Canada border—a frequent concern among his many American patients—he placated officials in letters, stating in one case that seizures were "merely symptoms of an old [brain] injury, not epilepsy."[108] He also preemptively advised patients and their families that "if some other word is used, instead of the word epilepsy, there is no difficulty whatsoever."[109] As he explained to Jean's father in 1947: "The immigration people both Canadian and American will refuse passage across the border to anyone who is labeled as an epileptic. Nevertheless, hundreds of thousands cross the border. . . . If no mention is made of this tendency, there is never any difficulty."[110] Despite his genuine conviction that epilepsy was wrongly stigmatized, he frequently told patients that it was better not to invoke the word at all.[111]

Yearly updates, solicited by the MNI to build postoperative data, offered patients further occasion to voice opinions and concerns about their surgical outcomes. Many though not all who returned forms were pleased or satisfied. In 1953 one patient ecstatically reported his "PERFECT HEALTH" a year after his surgery, adding "thank god and Wilder Penfield."[112] Another confessed each year into the 1970s that she never again experienced attacks but only occasional auras.[113] Others went into greater detail, describing what surgery had meant for their daily lives. As one husband of a Montreal patient noted five years after his wife's right temporal lobectomy, "She used to stay at home and stay away from people; now she goes out and has many friends." Another patient married shortly after gaining control of her seizures, and it was this event that demonstrated a "completely successful outcome."[114]

Success, however, was not defined uniformly by patients or their doctors at the MNI. According to Penfield and his colleagues, a "successful outcome" included those surgeries resulting in a quantifiable seizure reduction of 50 percent or more.[115] Patients' expectations were often more qualitative and personal in nature and could therefore differ. Those dissatisfied, or who deemed their improvement inadequate, sometimes expressed these sentiments in follow-up surveys. When asked "Are you sorry or glad you had the operation?" some patients reported that they were sorry. One patient, for instance, was regretful because the surgery "did not correct the difficulty."[116] Another from Buffalo, New York, said he regretted his 1933 operation because his driver's license had since been revoked and he could no longer work.[117] Others were dissatisfied with the outcome but nevertheless content with the decision to undergo surgery, remarking that "at least I tried."[118] Upon recurrence or a lack of reduction in their seizures, the choices available to patients were generally "new medicines" or sometimes further operations, as with Jean in 1954.[119] After dealing with the aftermath of surgery, however, patients rarely embraced such options enthusiastically.

Beyond inadequate improvement, surgeries also presented risks of death and impairment. Some patients reported postoperative memory loss, issues regarding fluidity of speech and "swiftness of thought," and other neurological "deficits."[120] Some committed suicide following their surgeries, experienced prolonged depression, or developed psychosis.[121] Others died during seizures, sometimes by drowning, crashing their cars, or falling down stairs, while for others, fatal complications followed surgery by days or weeks.[122] In these unfortunate cases, it was not unusual for families of the deceased to express regret about the decision to operate. The mother of the Vermont man who died two weeks after surgery of a wound infection wrote, "I cannot help but wonder if we took the right course of action or not."[123] Raising the fact that her son's seizures were "not always so bad or frequent," she nevertheless resolved that it had been worth it since "every time he had an attack we were nearly beside ourselves. . . . If there was a chance for him to be helped we wanted him to have it."[124] Like another patient's wish that the surgery would cure or kill, both happened, and so did everything in between.

Penfield, while confident in his abilities, was sometimes conflicted about his position of power and the trust that patients placed in him. In 1936, the same year that he first operated on Jean, several of his patients died. Speaking of the death of one Montreal patient, he wrote in his diary, "[Patient] is dead. I took a scar from his brain. I watched the brain while he had a fit—and bacteria—either introduced by his accident two years before, or by our careless technique—has closed his chapter. His wife, with a baby of two months, thanks me for what?"[125] Of the Vermont farmer, Penfield similarly lamented,

"I watched his brain while he had his fits without knowing why or what was happening. He used to tell me how much he appreciated the time I gave him. . . . I've been haunted by him at night."[126] Penfield to some degree tempered this inner conflict with "awe" of his patients' bravery and faith in him as a surgeon. To a group of medical students in 1947, he marveled, "An operator can only be filled with amazement at the heroism of men, women, and children who allow him his slender stock of skill to expose the brain to deal with it according to his discretion."[127] In Penfield's opinion, "the stoicism, perhaps I should say heroism, of this series of patients who have undergone these procedures with fortitude must make the neurosurgeon feel humble."[128]

Speaking Secrets

Wilder Penfield has been widely described as a brain explorer, his cortical travels yielding some of the first and most accurate functional maps of the human brain. Such remembrances and metaphors, however, tend to relegate his patients with epilepsy to the status of ambiguous terrain, obscuring their role in science, hiding their "stoicism," and minimizing the experiences that brought them to him in the first place. Only through efforts to treat intractable epilepsy did the justification emerge to stimulate and operate upon the seizure-prone brain; yet, if clinical aims aided access, it was patients whose participation and input made investigation meaningful. Where the brain was a mysterious but suddenly navigable "object" of science, the patient became a ready subject. Such conditions made Penfield's findings possible and the brain seemingly knowable in the mid–twentieth century.

Penfield in fact believed that his patients could "speak secrets." To varying degrees, patients—and in particular, his "record-of-past-experience" patients—offered entry into otherwise unreachable thoughts and sensations during their operations. Expanding feedback beyond the surgery itself, others provided information through surveys and letters over the course of many years. Jean, although somewhat unusual in the importance of her case, is one such patient. Penfield wrote to and of Jean, who was to him both medium and muse, frequently throughout his career. The two exchanged letters until Penfield's death in 1976, and she became the textbook example of what he termed "psychical" or "dreamy state" seizures. Given the psychological and emotional dimension of her auras, she was also the first patient in which memory, and therefore mind, was activated via stimulation—a topic that would animate Penfield's scientific and spiritual questioning for the remainder of his life and for which he frequently sought her elucidation.

Others from this era who underwent surgeries for epilepsy similarly saw scenes and heard fragments from their pasts in the operating room: popular songs from youth, childhood spaces, and visions of themselves giving birth. Importantly, each of these moments was shared rather than found; only through descriptions of what patients felt and thought in those moments could a "record of experience" be extrapolated from electrical stimulation. Equally, only through the patient's cooperative recounting of hearing a song or reliving a frightening experience could ideas about the physical action of memory be adequately developed. Theories regarding the brain and mind, once left to hypothesis and projections based on gross anatomy, became the interactive domain of surgeon and patient. Patients' participation and knowledge are, nevertheless, typically missing in the historical retelling.

Ultimately, although some of Penfield's patients acknowledged their role in science, the primary motivation of each had been to find relief from their seizures. Both the vast limitations surrounding epilepsy and faith in surgery as a curative force meant that many chose to operate to address conditions they assumed were incompatible with "normal life." Interactions between surgeon and patients, while relatively cooperative and collaborative, also remained uneven; at the physician's discretion, patients' contributions could be "reliable" and "insightful" or have "no bearing on anything." These divergences could extend to patients' experiences and understandings of their cases as well. While Jean traced her seizures to a personally meaningful moment, Penfield searched her medical history for evidence of prior illness, trauma, and sick relatives. Similarly, patients often differed in their definitions of a good surgical outcome. While Penfield might deem any overall reduction of seizures a "success," patients and family members frequently emphasized personal achievements—or, given policies at borders and in workplaces, were satisfied with no less than "complete control."

In 1960, weeks before her wedding and Penfield's retirement, Jean traveled to the MNI for a final time. In a photograph the two had taken together, Jean links arms with Penfield and smiles at the camera. She is thirty-eight years old and six years seizure free, and she considers hers a success story based on her newfound feelings of control. Captured in the frame is also something of the manifold secrets between doctor and patient. Reflecting on his life's work, Penfield often expressed solidarity and shared credit with his teachers, colleagues, and benefactors at the Institute. The inscribed copy of the photograph that he sent to Jean privately acknowledged a different debt: "To [Jean]," he wrote on the picture's surface, "whose gallant courage made many things possible." In many ways, this was the company to which early exploration of the brain was most clearly, albeit quietly, indebted.

Control Paradigms and the Invisible Ideal

When my sister was recovering from her second neurosurgery in 2001, we were both most of the way through high school and hoped that the procedure would control, if not stop, her seizures entirely. In the years leading up to those surgeries, her epilepsy had intensified to a life-threatening degree. For myself and our family, seizures were the stressful events I note at the beginning of this book—episodes that could approach medical emergencies in their length and frequency—and which over the years had rarely ceased to alarm us, however present they became. Yet Gillian's epilepsy had simultaneously seemed quite ordinary; only in the presence of strangers did seizures assume the added meaning of being extraordinary events. Gillian's own perspective, however, remains fundamentally different from mine. Irreducible to moments of management or response, epilepsy and seizures for her represented both a clear mark of her difference and a continuous source of restriction. Most days, she says, epilepsy was everything that she could not do—or, more accurately, was prevented from doing—in the name of her personal safety.[1]

Throughout this project, I have struggled with what it means to control and attempt to eliminate seizures. It is a goal that at once seems to further medicalize and stigmatize epilepsy, yet it is something that people with epilepsy have themselves wanted and pursued throughout this history for reasons beyond simply wishing to conform to a society that fails to tolerate seizures. Their reasons, though not always well documented, have included survival, personal health and safety, greater independence, privacy, access to different kinds of activities or jobs, convenience, and the peace of mind of knowing that a seizure will not or is less likely to occur.[2] On the one hand, the desire for autonomy and independence (or indeed, certainty), like the desire for "control," suggests a world that narrowly prioritizes individual self-sufficiency and composure at the expense of more expansive models of community acceptance or accountability. As disability scholars note, self-mastery and control are also largely illusory, given that most people experience disability at some point in their lives.[3] On the other hand, the wish to know and rely on one's body insofar as possible seems understandable and quite human. Indeed, as medical approaches to epilepsy in more recent decades have moved from tools that are preventative and reparative to those that are also interactive and predictive (including a range of wearable seizure-alert and even brain-computer interface technologies), the focus for experts, patients, and their families has still been to limit, or at least mitigate, the impact of seizures. As a 2021 *New Yorker* article on experimental neural implants revealed, interviewees who could not stop their seizures outright were often satisfied with an implant that gave advanced warning. Such warning offered them freedom to remove themselves from certain situations before seizures began.[4] Participants thus gained control over the "where" and "how" of their seizures, even if power over the "what" itself continued to elude them. As the article said of one woman, "She had never had a self that she could trust before."[5]

This book centrally argues, however, that issues of control and self-trust are neither neutral nor without contexts and histories of their own. While the concepts of control and self-trust are experienced and to some extent embodied, they also fundamentally bear a particular set of values. Likewise, they are inextricable from questions of *social* trust, as well as from external measures and perceptions of control that continue to shape individual experiences.

In the decades immediately following World War II, epilepsy was for the first time presented as a "controlled" illness and the "epileptic" as an essentially "safe" person, deserving of the rights and privileges of other citizens. Much of this transition relied on the promise of reducing seizures. Particularly with the development of new drugs that lessened seizure severity and incidence, physicians and early advocacy groups came to describe seizures as atypical interruptions rather than runaway occurrences: as episodes apart from the individual and daily life rather than a central characteristic or social marker. Simultaneously, epilepsy's champions often justified greater inclusivity on the grounds of treatability and control. Epilepsy was a condition that when medically managed produced people "like others."[6] The idea that medication and surgeries further normalized those with epilepsy helped condemn the "dead but odiferous laws" that persisted into the postwar decades.[7] What seemed to make formal and informal prohibitions against marriage, driving a car, or going to school increasingly unjust, however, was the seizure-prone person's assured normality as evidenced by control itself.

Changing medical and social expectations surrounding epilepsy mid-century thereby engendered new, and largely unseen, kinds of work. Although twentieth-century proponents of seizure control believed their efforts would foster the very opportunities that groups like the Laymen's League Against Epilepsy first championed in the early 1940s, they in other ways reflected new pressures. With the earliest suggestion that epilepsy was controllable came assertions of "what an epileptic can do for himself"—or indeed, what people with epilepsy *should* do for themselves.[8] Given the misalignments between available treatments and limited inclusivity, it effectively became the role of "near-normal" people with "part-time handicaps" to *control*, *contain*, and *conceal* their epilepsy.[9] These were the realities, if not the burdens, of social safety and belonging that continue to characterize many invisible illnesses and disabilities even today.

Control: Questions of Trust and Responsibility

By the 1950s, seizure control popularly referred to the seizure-suppressing effects of anticonvulsant drugs like Dilantin.[10] What the phrase obscured, however, were two important facts: one, that seizures for many persisted with treatment, and two, that the power to exert control was believed to reside almost exclusively with individuals. As one source highlighted in 1944, "Even with the best medical care the patient must want and work to get well."[11] It continued: "There may be some illnesses in which no harm results from leaving off medication for a day, or several days, but epilepsy is not one of them."[12] Patient manuals and advice literature of the postwar period similarly emphasized the lack of room for "patient error" and the need for a diligent approach.[13] Though presented as a medical outcome, control was thus as much an ethic of model patient behavior, one that simultaneously masked the considerable elements of chance and good fortune that characterized any "successful" treatment.

Control also required more than a resolute adherence to drug therapy or, for those who underwent surgery, postoperative care. In addition to taking medication regularly, seizure-prone people were encouraged to "do for themselves," including eating "the right kind of food" and sleeping "a normal number of hours."[14] Even though doctors deemphasized the connections between diet and seizures by the 1940s, they still saw food, rest, and sobriety as keys to self-regulation—as demonstrated, most notably, by those seeking driver's licenses. Moreover, whereas seizure-prone patients had previously been warned to avoid the numerous physically "exciting causes" of epilepsy, now they were encouraged to "keep busy" and, like student athletes in Detroit, to "build up [their] muscles by all sorts of exercise."[15] Equally, people who practiced control were curious and well versed in the science of epilepsy. This allowed individuals greater agency in their own self-care. As one source noted, "He should inform himself about his disorder so that he can work more intelligently to cure himself."[16] Curing, or at least treating oneself, included knowing the results of diagnostic tests and the names, dosages, and chemical components of one's medications. It also meant reading medical literature, keeping regular doctor's appointments, and recording all seizures. Rather than testimony that "bore no relation to anything," good patients presented the right information at the right time.[17] Given new knowledge about the importance of electroencephalography in determining one's soundness to marry or have children, those with epilepsy consulted with their doctors on this and most other major life events.[18] These changes made control seem nearer than ever before while also imposing rigid and even impossible new criteria of treatment.

Correspondingly, people with epilepsy were deemed increasingly responsible when seizures happened. Soon after the diagnostic and therapeutic developments of the 1930s, a seizure became a "failure"—either to "prevent" or to "fortify the seizure dam"—or simply evidence of noncompliance.[19] When children at the White Special School had seizures on school trips, or licensed drivers had seizure-related accidents, authorities and administrators often attributed such incidents to poor adherence to medical treatment. Whereas public reactions to the involuntary nature of one man's seizures in a small midwestern community of the 1920s had been characterized by resignation (people were somewhat "nice about it, as long as they knew he couldn't help it"), those who had seizures in schools, workplaces, or on public streets after this time "neglected" to take their medication or to get "enough rest."[20] If drugs and proper behaviors controlled epilepsy, then, in some basic way, only people with epilepsy could fail to control their own illnesses. One pamphlet captured these invisible new standards: "How big a handicap it is depends a good deal on how big a handicap he allows it to be."[21]

Contain: Conformity and Self-Management

Given that epilepsy was considered one of "the most individualized of disorders" in its presentation by the 1940s, it is fitting, perhaps, that its management also remained so personal in this era.[22] Part of the contract of control was to avoid unduly or unnecessarily taxing others with any aspect of one's condition. Doctors and advocates who promoted college education for people with epilepsy in the 1950s, for example, were quick to suggest that "colleges are not hospitals" and that an applicant should opt for private rather than dorm-style accommodation.[23] Insofar as possible, epilepsy was an intermittent illness held in suspense, with any symptoms or difficulties privately attended to. It was also vital for those with epilepsy to show little if any of the effort involved in practicing control. While there was room for evidence of these facts in a person's private quarters and medicine cabinet, they should not spill into public view.

Containing epilepsy in this way also meant that those with epilepsy needed to be stoic and minimize what their seizures entailed. In the postwar period, schoolchildren were encouraged to rise following a seizure "as if nothing had happened," while vocational rehabilitation literature claimed that the rare employee who had a seizure on the job took only a short break before resuming work. Described as "not a continuously rough sea" but a gentler recurrent wave, epilepsy became a collection of sporadic and barely indisposing episodes.[24] People with epilepsy were similarly encouraged to brace for the side effects of seizures and medications and, if eligible, to accept the risks associated with more radical procedures.[25] Moreover, when adversity did occur, they should not let it affect them, since seizure-prone people were "normal" and could dismiss any lingering public ignorance as just that. At the same time, it was important to be sensible, taking care not to impose danger or discomfort on others. Knowing one's "limits" included paying attention to seizure triggers and any auras preceding the onset of a seizure. Activities such as swimming or bathing unsupervised and "forbidden occupations" like working an elevator were also to be avoided. Though the mechanization of farm labor had made "pastoral work," once suggested for people with epilepsy, "as dangerous as the factory," other formerly prohibited practices, like walking alone, having children, or driving a car, became safe and even recommended.[26]

Above all, to be a conscientious seizure-prone person meant adhering to social norms and expectations wherever possible. By the late postwar period in particular, advocates suggested that the "real handicap" of epilepsy was "more social than physical."[27] Such arguments foregrounded what people with epilepsy could do to address the social challenges they faced rather than on what society itself might do to change. Being steadily employed, being part of a family unit, and showing few traces of one's disability constituted the right approach.[28] "He can follow the news and discuss it," one source suggested. "He can become interested in politics . . . [and] have a hobby."[29] Well-rounded and publicly attuned, the person with epilepsy was to "consider himself a perfectly healthy, normal human being—which he may be except during his seizures—and live like one."[30] After World War II, popular depictions of epilepsy thus frequently featured students and professionals whose seizures were minor occurrences in their lives. While such sources acknowledged that those with epilepsy "had a handicap to be sure," they also noted that "so have most other persons" and moreover that epilepsy "may disappear very quickly under the right treatment."[31] In addition to not dwelling on one's illness, people with epilepsy were advised against engaging in self-pity or seeking "special [social] treatment." The child with epilepsy who reached her full "potential" therefore needed no additional assistance in school or beyond. Patients "cured" through surgery could embrace marriage, work, and the responsibility of showing just how "normal" they were.

Conceal: Fragmentation and the Invisible Ideal

Although legacies of shame and secrecy around epilepsy were increasingly deemed "cankers," to which openness and honesty became the "salve," hiding one's epilepsy also seems to have been an integral part of living with the condition after World War II.[32] Some contexts demanded that people tell the truth about their epilepsy unequivocally, yet there were many others in which such transparency was unwise and even unwelcome. Doctors, family members, and religious leaders remained proper confidants, whereas neighbors, coworkers, and acquaintances were rife with risk.[33] Physicians at the time debated whether it was prudent for patients to inform employers and school administrators of their epilepsy. Those in favor recommended offsetting the news with evidence of a superior work ethic.[34] Revealing epilepsy was similarly advised in all matters of public safety. To misinform or fail to mention a propensity for seizures to a public official, in the army, or at a driver's licensing bureau was not only dishonest, it was also dangerous.

At the same time, however, it was tacitly understood in this era that people should refrain from discussing their epilepsy publicly. It was not socially appropriate, for example, to casually inform a stranger of one's epilepsy—even when it was a truthful answer to a direct question. As before, those who did were often censured or dismissed.[35] The ability to leave one's epilepsy unstated, in fact, became a sort of invisible ideal. Dr. Tracy Putnam hinted at the virtues and hidden advantages of this approach in 1943, stating, "With rare exceptions, he [with seizures] would pass unnoticed in the throng of mankind."[36] Certainly, obscuring one's epilepsy had always existed as an act of strategic omission. This is evident in early twentieth-century examples of people seeking admission to schools and at borders. Nevertheless, mid-century beliefs regarding the seizure-prone person's treatability and near normalcy also promoted these behaviors. Redefined as a respectably "quiescent" condition, epilepsy was far from brought to public light.

As a consequence, seizures retained their power to dismantle any positive consensus around a seemingly intermittent illness, exposing "normal" people as "epileptic" and raising alarm about their tenuous relationship to control. As with other less visible illnesses and disabilities, tensions invariably arose when appearances faltered.[37] Throughout the postwar era, seizures continued to induce fear and those with "known" epilepsy faced ongoing exclusion. Parents often complained that others refused their children as playmates. Many other seizure-prone people were fired from their jobs. Intolerance was also clear in what several described as the "persistent dread" that characterized their lived experience of epilepsy, both before and after the rise of anticonvulsant medications and new surgeries.[38] According to Dr. William Lennox, public seizures bore particularly devastating effects for the individual: "A healthy person, happy and productive, is suddenly bereft of awareness, of control of body and mind, a fearful spectacle to all who behold him."[39]

In addition, not everyone was equally advantaged by measures of seizure control. Those who had frequent seizures, who had less "productive potential," or who lacked access to regular medical care encountered additional barriers. Like students whose remaining disabilities kept them at Detroit's school for children with epilepsy—or whose race and class determined which schools they attended next—becoming a "near-normal" person with a "part-time handicap" was never assured or uniformly achieved.[40] Others were denied the physical and social mobility afforded by a driver's license owing to their gender, marital status, or ostensibly questionable character.[41] For still others, an electroencephalographic reading meant they would forever have to disclose their cerebral dysrhythmias to certain employers, officials, or potential partners.[42] For these individuals, as well as for the estimated 10–20 percent of people who saw no improvement with medication or who were placed in institutional care, the nature of these postwar medical and social developments was more ambiguous.

Invisibility and Risk

There have been changes since the 1960s not addressed by this book. The civil rights movement, women's and gay liberation, critiques of professional medicine, deinstitutionalization, and the disability rights movement itself all had an impact on the culture and course of this history. Each in different ways foregrounded issues of social oppression, discrimination, and identity. Each demanded more open and inclusive discourse and policy.[43] By the 1960s and 1970s, children with epilepsy and other disabilities increasingly attended regular schools, while remaining laws against marriage, immigration, and sterilization were overturned state by state.[44] Augmenting parents' and patients' groups, the Epilepsy Foundation, founded in 1968, became a national source of information and advocacy in the United States. Disability organizations across the country also fought for rights-based measures, culminating most notably in the Americans with Disabilities Act (ADA) in 1990. In medicine, a number of developments, including CAT scans, MRI technology, online EEG monitoring, and improved drugs and surgeries, continued to expand the diagnostic and treatment options available.[45] Yet even today, almost a century after the formation of the Harvard Epilepsy Commission, epilepsy is still described by experts as among the most intractable and misunderstood of conditions.

Many of the forms of management traced by this book also continue. By and large, epilepsy remains an "invisible" neurological disorder and disability, often described as imperceptible but challenging to navigate. Stigma and public misconceptions about epilepsy correspondingly persist. Much educational literature to the present, including resources for children and parents, addresses misinformation and stereotypes.[46] In addition to the apparent need for myth busting, the rhetoric of being "no different" from others, often with the caveat of "except for seizures," also remains. Simultaneously, people with epilepsy continue to cite feelings of alienation they say foster forms of social masking.[47] Although the ADA protects against discrimination on the basis of disability—meaning, for example, that people with epilepsy cannot be fired from their jobs or denied promotions because of their epilepsy—many still describe what postwar advocates once called the "epileptic's dilemma" of disclosure, or "Should he tell?"[48] It remains difficult to know the extent to which negative preconceptions in a variety of social situations might prevail. Placing trust in the relative "invisibility" of one's condition may in some cases seem, or indeed be, safer.

Within histories of medicine, discussions of risk usually focus on perils of a physical nature: the chance of getting sick or becoming sicker, the potential danger posed by certain therapies, and the measures taken to assess and assuage these real and apparent threats.[49] In contrast, the history of epilepsy since the mid–twentieth century is as much a history of social risk. As bearers of an intermittent, invisible, and newly public disability in the post–World War II era, seizure-prone individuals encountered risks that impacted their personal relationships and navigation of public space. The burden was—and in many cases has remained—on individuals to determine how big a disability their epilepsy might appear to be. The origin of these frameworks can arguably be found in an era of incomplete medical control and a rising invisible ideal, born of an enduring taboo against seizures.

Control, and the self-management required in its service, redefined what it meant to have epilepsy in the twentieth century. Mid-century medical innovations did not produce changes to seizure-prone people in expected or even consistent ways; they did, however, promote new strategies for managing epilepsy's uncertainties and risks. By exploring seizure control and invisible disability, this book shows both what is at stake in losing control of one's body and to what ends people are compelled to calculate and to perform diverse measures of control. This contributes to a larger story of how notions of health and able-bodiedness pervade experiences of chronic illness and disability more broadly—often in the name of personal independence and autonomy—and the extent to which new ways of understanding or being might be possible.

Notes

Introduction

1. All patients and those who wrote to Dr. Penfield about their medical concerns are anonymized here and throughout the book. Letter A to Dr. Wilder Penfield, May 1949, box 44, C/G 7, Wilder Penfield Fonds, Osler Library for the History of Medicine (OLHM), McGill University, Montreal, QC.

2. Letter B to Penfield, Dec. 1948, box 44, C/G 7, OLHM.

3. Letter C and Letter D to Penfield, Apr. 1948, box 44, C/G 7, OLHM.

4. Patients' letters to doctors, whether seeking treatment or coming through doctor-patient organizations like the Laymen's League Against Epilepsy, provide one of the few windows into experiences of epilepsy in this period.

5. On stigma, eugenics, and epilepsy in the early twentieth century: Ellen Dwyer, "Stigma and Epilepsy," *Transactions and Studies of the College of Physicians of Philadelphia,* 13, no. 4 (1991): 387–410.

6. Paul Lombardo, *Three Generations, No Imbeciles: Eugenics, the Supreme Court, and Buck v. Bell* (Baltimore: Johns Hopkins University Press, 2008).

7. On school exclusions: Hildred Ileene Jarvis, "A Study of the Detroit Experimental School for Epileptic Children" (MA diss., Institute of Public and Social Administration, University of Michigan, 1937). On immigration and epilepsy: Douglas Baynton, "Defectives in the Land: Disability and American Immigration Policy," *Journal of American Ethnic History* 24, no. 3 (2005): 31–44, 33. On the rise of epileptic farms and colonies, especially the Craig Colony in New York: Ellen Dwyer, "Stories of Epilepsy, 1880–1930," in *Framing Disease: Studies in Cultural History,* ed. Charles Rosenberg and Janet Golden (New Brunswick: Rutgers University Press, 1992).

8. Such characterizations and definitions of "normal" appeared in many news stories and public sources. Among the earliest is Herbert Yahraes, *Epilepsy—The Ghost Is out of the Closet,* Public Affairs Pamphlet no. 98 (New York: Public Affairs Committee, 1944), 3–4.

9. Dilantin, EEG, and neurosurgical advances of the mid-1930s are often cited as leading to new forms of inclusion and public participation for people with epilepsy. Lewis P. Rowland, *The Legacy of Tracy J. Putnam and H. Houston Merritt: Modern Neurology in the United States* (New York: Oxford University Press, 2008); Milton Silverman, "We Can Lick Epilepsy," *Saturday Evening Post,* Jan. 17, 1948, 107–14; David Anderson and Robert Cromie, "Epilepsy Isn't a Curse!" *Better Homes and Gardens,* March 1948, 180–85.

10. For example: William Lennox, "Epilepsy: The Hopeful Disorder," *Parents' Magazine,* Feb. 1946, 32; "Some Plain English on Epilepsy," *Science Digest* 17 (1945), 82–83.

11. For example: Allan Brandt and Martha Gardner, "The Golden Age of Medicine?" in *Companion to Medicine in the Twentieth Century,* ed. Roger Cooter and John Pickstone (New York: Routledge, 2000); Bert Hansen, *Picturing Medical Progress from Pasteur to Polio:*

A History of Mass Media Images and Popular Attitudes in America (New Brunswick: Rutgers University Press, 2009).

12. This statistic appeared in several sources, including news and vocational rehabilitation materials, by the mid-1950s. For instance: Frederic Gibbs and Frederick Stamps, *Epilepsy Handbook* (Springfield, IL: Charles C. Thomas, 1958), 3–4. By 1960, experts citing a Merritt-Putnam series stated that 85 percent of people with epilepsy reduced the number of seizures they experienced significantly with available medications: William G. Lennox, *Epilepsy and Related Disorders* (Boston: Little, Brown, 1960), 890. In the 1980s, the same statistic was cited by sociologists Joseph Schneider and Peter Conrad, who noted the ways in which "cure" had remained rare whereas "control" had not: Joseph W. Schneider and Peter Conrad, *Having Epilepsy: The Experience of Control and Illness* (Philadelphia: Temple University Press, 1983), 51.

13. "Who Can Get Epilepsy?," Epilepsy Foundation, Feb. 4, 2022, https://www.epilepsy.com/what-is-epilepsy/understanding-seizures/who-gets-epilepsy; "What Is Epilepsy/What Are Seizures?," Epilepsy Foundation, revised by Joseph I. Sirven and Patricia O. Shafer, Mar. 2014, accessed Oct. 14, 2014, www.epilepsy.com.

14. "Epilepsy and Seizures," National Institute of Neurological Disorders and Stroke, revised January 23, 2025, https://www.ninds.nih.gov/health-information/disorders/epilepsy-and-seizures.

15. This describes, in order, what are currently called tonic-clonic, absence, and myoclonic seizures.

16. Sudden, early, and unexpected death affects approximately 1 in 1,000 people with epilepsy and typically occurs among those with uncontrolled or drug-resistant epilepsy. "SUDEP," Epilepsy Foundation, revised September 21, 2022, https://www.epilepsy.com/complications-risks/early-death-sudep.

17. "Epilepsy and Seizures," National Institute of Neurological Disorders and Stroke; John Paul Leach and Rebecca O'Dwyer, *Epilepsy Simplified* (Malta: Gutenberg Press, 2011), xii, 1, 4; Orrin Devinksy, *A Guide to Understanding and Living with Epilepsy* (Philadelphia: F. A. Davis, 1994), 235–46.

18. Schneider and Conrad, *Having Epilepsy*, 104.

19. Schneider and Conrad, *Having Epilepsy*, 104–7.

20. Margiad Evans, *A Ray of Darkness* (New York: Roy Publishers, 1953), 122–24.

21. On control more generally in the history of medicine, including therapeutic and preventative forms: Christopher Feudtner, *Bittersweet: Diabetes, Insulin, and the Transformation of Illness* (Chapel Hill: University of North Carolina Press, 2003); Thomas Schlich, "Negotiating Technologies of Surgery: The Controversy about Surgical Gloves in the 1890s," *Bulletin of the History of Medicine* 87, no.2 (2013): 170–97. On the myth of medical control: Susan Wendell, *The Rejected Body: Feminist Philosophical Reflections on Disability* (New York: Routledge, 1996), 85–96.

22. On disability, productivity, and social citizenship: Sarah Rose, *No Right to Be Idle: The Invention of Disability, 1840s–1930s* (Chapel Hill: University of North Carolina Press, 2017). Questions of control and social legibility are also closely related to those of efficiency and progress in the twentieth century: Jennifer Karns, *The Mantra of Efficiency: From Water Wheel to Social Control* (Baltimore: Johns Hopkins University Press, 2008).

23. Risk in the history of medicine usually refers to epidemiological risks, risks associated with new procedures, or the way being at risk for a disease moves more people into the category of the disease itself: Robert Aronowitz, "The Converged Experience of Risk and Disease," *Milbank Quarterly* 87, no. 2 (2009): 417–42; Thomas Schlich and Ulrich Tohler, *The Risks of Medical Innovation: Risk Perception and Assessment in Historical Contexts* (New York: Routledge, 2005).

24. On models of self-care and the internalization of control: Michel Foucault, *The Birth of Biopolitics: Lectures at the Collège de France, 1978–79,* ed. Michel Senellart (Basingstoke: Palgrave Macmillan, 2008); Michel Foucault, *Discipline and Punish: The Birth of the Prison* (New York: Vintage Books, 1995).

25. For early examples of epilepsy as "invisible" in popular and private sources: Steven M. Spencer, "No Wonder Epileptics Are Bitter," *Saturday Evening Post*, Mar. 28, 1953; Patient E to Penfield, Dec. 1948, box 44, C/G 7, OLHM.

26. Patient F to Penfield, Dec. 1948, box 44, C/G 7, OLHM.

27. For instance: Patient F to Penfield; Spencer, "No Wonder Epileptics Are Bitter."

28. Peter Conrad and Joseph W. Schneider, eds., *Deviance and Medicalization: From Badness to Sickness* (Philadelphia: Temple University Press, 1992). On stigma as it pertains to disease, disability, and often identity: Erving Goffman, *Stigma: Notes on the Management of Spoiled Identity* (New York: Jason Aronson, 1974); Susan Sontag, *Illness and Metaphor and AIDS and Its Metaphors* (New York: Doubleday, 1988); Allan Brant, *No Magic Bullet: A Social History of Venereal Disease in the United States since 1880* (New York: Oxford University Press, 1985).

29. Schneider and Conrad, *Having Epilepsy*, 16.

30. Schneider and Conrad, *Having Epilepsy*, 16, 22.

31. Schneider and Conrad, *Having Epilepsy*, xi–xii.

32. Some sources beginning in the 1940s reported that as much as 1 percent of the US population had epilepsy and that fewer than 10 percent of people with epilepsy lived in institutional settings: William Lennox, *Epilepsy Explained* (Boston: Laymen's League Against Epilepsy, 1940). Schneider and Conrad's research in the late 1970s suggests a similar percentage; they note that approximately 50,000 Americans with epilepsy lived in institutional care in the 1950s: Schneider and Conrad, *Having Epilepsy*, 27.

33. As many noted to their doctors throughout the period, revealing epilepsy was risky in a variety of legal, financial, and interpersonal ways. See patient letters and records of Wilder Penfield, Montreal Neurological Institute-Hospital (MNI), Montreal, QC. On twentieth-century privacy and the relationship between public and private selves: Sarah Igo, *The Known Citizen: A History of Privacy in Modern America* (Cambridge: Harvard University Press, 2018).

34. On new and old laws related to epilepsy in this era: Gerald Walker, "Epilepsy: Medical Triumph, Social Tragedy," *Pageant*, Dec. 1960, 93–97.

35. For epilepsy's history, in chronological order of focus: Owsei Temkin, *The Falling Sickness: A History of Epilepsy from the Greeks to the Beginnings of Modern Neurology* (Baltimore: Johns Hopkins University Press, 1945); Dea Boster, "An 'Epileptik' Bondswoman: Fits, Slavery, and Power in the Antebellum South," *Bulletin of the History of Medicine* 83, no. 2 (2009): 271–301; M. J. Eadie, *A Disease Once Sacred: A History of the Medical Under-*

standing of Epilepsy (London: John Libbey, 2001); Walter Friedlander, *A History of Modern Epilepsy: The Beginning, 1865–1914* (Westport, CT: Greenwood, 2001); Dwyer, "Stories of Epilepsy"; Dwyer, "Stigma and Epilepsy"; Ellen Dwyer, "Neurological Patients as Experimental Subjects," in *The Neurological Patient in History,* ed. Stephen Jacyna and Stephen Casper (Rochester, NY: University of Rochester Press, 2012).

36. On medical developments for and around epilepsy: Eadie: *A Disease Once Sacred;* Friedlander, *A History of Modern Epilepsy;* Rowland, *The Legacy of Tracy J. Putnam and H. Houston Merritt.* On premodern views of epilepsy: Temkin, *The Falling Sickness.* On eugenic institutions for epilepsy: Dwyer, "Stories of Epilepsy" and "Stigma and Epilepsy."

37. Temkin, *The Falling Sickness*, 347; Schneider and Conrad, *Having Epilepsy*, 26–30.

38. Eadie, *A Disease Once Sacred;* Friedlander, *A History of Modern Epilepsy.*

39. Dwyer, "Stories of Epilepsy," "Stigma and Epilepsy," and "Neurological Patients as Experimental Subjects"; Boster, "An 'Epileptik' Bondswoman."

40. For a key work complicating the view of medicine as illness's logical resolution: Feudtner, *Bittersweet.* As noted, those histories of epilepsy focused on the late nineteenth- and early twentieth-century period often center on institutional settings. See Ellen Dwyer's critical work on the Craig Colony: "Stories of Epilepsy" and "Stigma and Epilepsy."

41. For example: Silverman, "We Can Lick Epilepsy," 109–10; Anderson and Cromie, "Epilepsy Isn't a Curse"; Rowland, *The Legacy of Tracy J. Putnam and H. Houston Merritt.*

42. Schneider and Conrad, *Having Epilepsy*, 29.

43. Feudtner, *Bittersweet*, 21–25; Robert Aronowitz, *Unnatural History: Breast Cancer and American Society* (New York: Cambridge University Press, 2007); Jose van Dijck, *The Transparent Body: A Cultural Analysis of Medical Imaging* (Seattle: University of Washington Press, 2005).

44. Schneider and Conrad's work on epilepsy was unique in this respect, focusing on the everyday experiences of people with epilepsy. The sociologists conducted 80 interviews between 1976 and 1979 with men and women ranging in age from 14 to 54 years old. Schneider and Conrad, *Having Epilepsy*, preface.

45. Tom Shakespeare, "The Social Model of Disability," in *The Disability Studies Reader,* 5th ed., ed. Lennard J. Davis (New York: Routledge, 2017), 196–97; Michael Oliver, *The Politics of Disablement* (London: Macmillan, 1990).

46. Some disability scholars have criticized the seemingly oppositional framework created by these models, including the way in which a social model may deny experiences of impairment. Shakespeare, "The Social Model," 197–202. Within the history of medicine and disability history, see Beth Linker, "On the Borderland of Medical and Disability History: A Survey of the Fields," *Bulletin of the History of Medicine* 87, no. 4 (2013): 499–535.

47. William Lennox "Rehabilitation of Epileptic Service Men," *American Journal of Psychiatry* 100 (1943): 202–4.

48. Rosemarie Garland Thomson, *Extraordinary Bodies: Figuring Disability in American Culture and Literature* (New York: Columbia University Press, 1997); Susan Schweik, *The Ugly Laws: Disability in Public* (New York: New York University Press, 2009); Paul K. Longmore, *Telethons: Spectacle, Disability, and the Business of Charity* (New York: Oxford University Press, 2016).

49. Susan Wendell, "Unhealthy Disabled: Treating Chronic Illness as Disabilities,"

Hypatia 16, no. 4 (2001): 17–33; Wendell, *The Rejected Body*; Ellen Samuels, "My Body, My Closet: Invisible Disability and the Limits of Coming Out Discourse," *Gay and Lesbian Quarterly* 9, nos. 1–2 (2003): 233–55; Ann Davis "Invisible Disability," *Ethics* 116, no. 1 (2005): 153–213; Jeffrey A. Brune and Daniel J. Wilson, eds., *Disability and Passing: Blurring the Lines of Identity* (Philadelphia: Temple University Press, 2013); Nancy Hirschmann, "Invisible Disability: Seeing, Being, Power," in *Civil Disabilities: Citizenship, Membership, and Belonging,* ed. Nancy Hirschmann and Beth Linker (Philadelphia: University of Pennsylvania Press, 2015).

50. Examples within the passing literature include physical disabilities from childhood polio, blindness, deafness, mental illness, cognitive disabilities, and menstruation: Brune and Wilson, *Disability and Passing,* 1–2, 10.

51. Wendell, *The Rejected Body*; Robert Aronowitz, "From Myalgic Encephalitis to Yuppie Flu: A History of Chronic Fatigue Syndromes," in *Framing Disease: Studies in Cultural History,* ed. Charles E. Rosenberg and Janet Golden (New Brunswick: Rutgers University Press, 1991); Meghan O'Rourke, *The Invisible Kingdom: Reimagining Chronic Illness* (New York: Riverhead Books, 2022).

52. For contemporary examples and theory: Wendell, "Unhealthy Disabled"; Samuels, "My Body, My Closet"; Ann Davis, "Invisible Disability"; Hirschmann, "Invisible Disability." While historical examples do appear in the Brune and Wilson volume, "invisibility" is not a state named by the actors involved in these particular cases: Brune and Wilson, *Disability and Passing*.

53. Peter Wolf, "Epilepsy in Literature," *Epilepsia* 36 (1995, Suppl. 1): S12–17, S12; Jeannette Stirling, *Representing Epilepsy: Myth and Matter* (Liverpool: Liverpool University Press, 2010); Jarvis, "A Study of the Detroit Experimental School."

54. Lennox, *Epilepsy Explained*; Waldemar Kaempffert, "Studies of Epilepsy: Review of *Science and Seizures,* by William G. Lennox," *New York Times,* Mar. 23, 1941.

55. Wendell, *The Rejected Body,* 85–96. Discussions of social citizenship in this book fall more in line with usages employed by disability scholars, particularly in reference to issues of access, inclusion, freedom from discrimination, as well as rights, representation, and belonging: Rose, *No Right to Be Idle,* 2–4; Douglas Baynton, "Disability and the Justification of Inequality in American History," in *The New Disability History: American Perspectives* (New York: New York University Press, 1999), 33–36; Alison Carey, *On the Margins of Citizenship: Intellectual Disability and Civil Rights in Twentieth-Century America* (Philadelphia: Temple University Press, 2009); Nancy Hirschmann and Beth Linker, "Disability, Citizenship, and Belonging: A Critical Introduction," in *Civil Disabilities: Citizenship, Membership, and Belonging,* ed. Nancy Hirschmann and Beth Linker (Philadelphia: University of Pennsylvania Press, 2015).

56. On disability, visibility, and social belonging: Thomson, *Extraordinary Bodies.* On disability, labor, and belonging: Rose, *No Right to Be Idle,* 2–4.

57. For instance: Mary L. Dudziak, *Cold War Civil Rights: Race and the Image of American Democracy* (Princeton: Princeton University Press, 2011); Elaine Tyler May, *Homeward Bound: American Families and the Cold War Era* (New York: Basic Books, 1988).

58. On parallel mid-century concerns about seemingly duplicitous persons and bodies: Jennifer Terry, *An American Obsession: Science, Medicine, and Homosexuality in*

Modern Society (Chicago: University of Chicago Press, 1999); Margot Canaday, *The Straight State: Sexuality and Citizenship in Twentieth-Century America* (Princeton: Princeton University Press, 2009); David K. Johnson, *The Lavender Scare: The Cold War Persecution of Gays and Lesbians in the Federal Government* (Chicago: University of Chicago Press, 2004).

59. John D'Emilio, *Sexual Politics, Sexual Communities: The Making of a Homosexual Minority in the United States, 1940–1970* (Chicago: University of Chicago Press, 1983). On the politics of respectability and race before the civil rights movement: Evelyn Brooks Higginbotham, *Righteous Discontent: The Women's Movement in the Black Baptist Church, 1880–1920* (Cambridge: Harvard University Press, 1994).

60. Overcoming narratives are a major theme identified by disability scholars and reflective of the medical model. For example: Emily Rapp, *Poster Child: A Memoir* (New York: Bloomsbury, 2006); Longmore, *Telethons*, 112–20.

61. On Cold War containments beyond those of national security: May, *Homeward Bound*; Jane Sherron De Hart, "Containment at Home: Gender, Sexuality, and National Identity in Cold War America," in *Rethinking Cold War Culture*, ed. Peter J. Kuznick and James Gilbert (Washington, DC: Smithsonian Institute Press, 2001); Arnold R. Hirsch, "Containment on the Homefront: Race and Federal Housing Policy from the New Deal to the Cold War," *Journal of Urban History* 26 (2000): 158–89.

62. Evans, *A Ray of Darkness*, 123.

63. Note that several disability groups organized around issues of employment, community transition, and education throughout the decades preceding the 1960s and 1970s, including notably the League of the Physically Handicapped, We Are Not Alone, and the National Association for Retarded Children. Paul K. Longmore and David Goldberger, "The League of the Physically Handicapped and the Great Depression: A Case Study in the New Disability History," *Journal of American History* 87, no. 3 (2000): 888–922.

64. For a typical example of the postwar child with epilepsy represented in US magazines: Matthias M. Krenn, "Social Outcast—Age 9," *Today's Health* 29, no. 8 (1951): 28–30, 30.

65. On mid-century normalcy: Anna Creadick, *Perfectly Average: The Pursuit of Normality in Postwar America* (Amherst: University of Massachusetts Press, 2010); Sarah Igo, *The Averaged American: Surveys, Citizens, and the Making of a Mass Public* (Cambridge: Harvard University Press, 2008).

66. Both patient and disability histories commonly center first-person narratives, though disability historians more often note institutional source issues and the ableism in the archive. Those who wrote about their experiences of epilepsy in the mid–twentieth century often used pseudonyms when contacting patient groups, magazines, and even doctors. For recent examples that more generally intersperse personal narratives of illness and disability within historical accounts: Emily K. Abel, *Sick and Tired: An Intimate History of Fatigue* (Chapel Hill: North Carolina University Press, 2021); Jaipreet Virdi, *Hearing Happiness: Deafness Cures in History* (Chicago: University of Chicago Press, 2020).

67. Thus far, Dea Boster and Anne Fadiman have interrogated the experiences of some Black, Asian, Indigenous, and immigrant seizure-prone people in more historically

distant and recent contexts: Boster, "An 'Epileptick' Bondswoman"; Anne Fadiman, *The Spirit Catches You and You Fall Down: A Hmong Child, Her American Doctors, and the Collision of Two Cultures* (New York: Farrar, Straus and Giroux, 1997).

Chapter 1. "Invisible Handicap"

1. Max Brand, *Dr. Kildare's Crisis* (New York: Grosset & Dunlap, 1940), 165; *Dr. Kildare's Crisis,* directed by Harold S. Bucquet (1940), UCLA Film & Television Archive, Archives Research and Study Center, Los Angeles, CA.

2. Steven M. Spencer, "No Wonder Epileptics Are Bitter," *Saturday Evening Post*, Mar. 28, 1953, 26.

3. Spencer, "No Wonder Epileptics Are Bitter," 26.

4. Spencer, "No Wonder Epileptics Are Bitter," 26.

5. On epilepsy, families, and institutionalization in the early twentieth century: Ellen Dwyer, "Stories of Epilepsy, 1880–1930," in *Framing Disease: Studies in Cultural History,* ed. Charles Rosenberg and Janet Golden (New Brunswick: Rutgers University Press, 1992).

6. On Dilantin and pharmaceutical control in 1938: Lewis P. Rowland, *The Legacy of Tracy J. Putnam and H. Houston Merritt: Modern Neurology in the United States* (New York: Oxford University Press, 2008).

7. On disability and passing: Jeffrey A. Brune and Daniel J. Wilson, eds., *Disability and Passing: Blurring the Lines of Identity* (Philadelphia: Temple University Press, 2013); Elaine K. Ginsburg, ed., *Passing and the Fictions of Identity* (Chapel Hill: Duke University Press, 1996), 1–18. On how the gaze of the "normate" constructs disability: Rosemarie Garland Thomson, *Extraordinary Bodies: Figuring Disability in American Culture and Literature* (New York: Columbia University Press, 1997); Susan Schweik, *The Ugly Laws: Disability in Public* (New York: New York University Press, 2009).

8. On invisible or less visible illnesses and disabilities: Susan Wendell, "Unhealthy Disabled: Treating Chronic Illness as Disabilities," *Hypatia* 16, no. 4 (2001): 17–33; Ann Davis "Invisible Disability," *Ethics* 116, no.1 (2005): 153–213; Meghan O'Rourke, *The Invisible Kingdom: Reimagining Chronic Illness* (New York: Riverhead Books, 2022); Nancy Hirschmann, "Invisible Disability: Seeing, Being, Power," in *Civil Disabilities: Citizenship, Membership, and Belonging*, ed. Nancy Hirschmann and Beth Linker (Philadelphia: University of Pennsylvania Press, 2015). For a comprehensive treatment of "invisibility" and the tension between appearance and identity: Ellen Samuels, "My Body, My Closet: Invisible Disability and the Limits of Coming-Out Discourse," *Gay and Lesbian Quarterly* 9, no. 1–2 (2003): 233–55.

9. On models of open secrecy during the World War II and postwar eras as examined by historians of sexuality: Heather Murray, *Not in This Family: Gays and the Meaning of Kinship in Postwar North America* (Philadelphia: University of Pennsylvania Press, 2010); Allan Berube, *Coming Out under Fire: The History of Gay Men and Women in World War II* (New York: Free Press, 2000).

10. The magazine serial version, "Dr. Kildare's Crisis," appeared in four parts in *Argosy* between December 21, 1940, and January 11, 1941. The film was released in December 1940. On representations of epilepsy in film and literature: Jeannette Stirling, *Representing Epilepsy: Myth and Matter* (Liverpool: Liverpool University Press, 2010); Jennie F.

Kerson et al., "The Depiction of Seizures in Film," *Epilepsia* 40, no. 8 (1999): 1163–67; Peter Wolf, "Epilepsy in Literature," *Epilepsia* 36, no. S1 (1995): S12–S17.

11. Motion Picture Association of America/Production Code Administration (MPAA/PCA), *Dr. Kildare's Crisis* Records, Margaret Herrick Library (MHL), Academy of Motion Picture Arts and Sciences, Beverly Hills, CA.

12. Faust only published his poetry under his real name: William A. Bloodworth Jr. See *Max Brand* (New York: Twayne, 1993). On the responses of servicemen to Brand's death: Darrell C. Richardson, ed., *Max Brand: The Man and His Work* (Los Angeles: Fantasy, 1952), 123–26.

13. Brand, *Dr. Kildare's Crisis*, 20.

14. Brand, *Dr. Kildare's Crisis*, 19.

15. Brand, *Dr. Kildare's Crisis*, 43.

16. Brand, *Dr. Kildare's Crisis*, 21.

17. Brand, *Dr. Kildare's Crisis*, 68, 127.

18. Brand, *Dr. Kildare's Crisis*, 36.

19. Brand, *Dr. Kildare's Crisis*, 36.

20. Brand, *Dr. Kildare's Crisis*, 79–80.

21. This is in keeping with earlier twentieth-century ideas about diet causing seizures and seizures being difficult to stop once underway; Brand, *Dr. Kildare's Crisis*, 144–47.

22. Charles Davenport and David Weeks, "A First Study of Inheritance of Epilepsy," *Journal of Nervous and Mental Disease* 38 (1911): 641–671.

23. On the conept of "healthy carriers" as it relates to infectious disease: Judith Walzer Leavitt, *Typhoid Mary: Captive to the Public's Health* (Boston: Beacon Press, 1996).

24. On popular and medical conceptions of epilepsy before 1930: Owsei Temkin, *The Falling Sickness: A History of Epilepsy from the Greeks to the Beginnings of Modern Neurology* (Baltimore: Johns Hopkins University Press, 1945); M. J. Eadie, *A Disease Once Sacred: A History of the Medical Understanding of Epilepsy* (London: John Libbey, 2001); Walter Friedlander, *A History of Modern Epilepsy: The Beginning, 1865–1914* (Westport, CT: Greenwood, 2001); Dwyer, "Stories of Epilepsy."

25. Brand, *Dr. Kildare's Crisis*, 74–75.

26. Brand, *Dr. Kildare's Crisis*, 146.

27. On popular, medical, and censor board reception to the film: MPAA/PCA, *Dr. Kildare's Crisis* Records, MHL.

28. On American eugenics and reproduction: Wendy Kline, *Building a Better Race: Gender, Sexuality, and Eugenics from the Turn of the Century to the Baby Boom* (Berkeley: University of California Press, 2005); Alexandra Stern, *Eugenic Nation: Faults and Frontiers of Better Breeding in Modern America* (Los Angeles: University of California Press, 2005); Daniel Kevles, *In the Name of Eugenics: Genetics and the Uses of Human Heredity* (Berkeley: University of California Press, 1985).

29. On race, class, eugenics, and institutionalization: Susan Burch and Hannah Joyner, *Unspeakable: The Story of Junius Wilson* (Chapel Hill: University of North Carolina Press, 2007); Nicole Hahn Rafter, *White Trash: The Eugenic Family Studies, 1877–1919* (Boston: Northeastern University Press, 1988).

30. Davenport and Weeks, "A First Study of Inheritance of Epilepsy," based on family

histories drawn from inmates at the New Jersey State Village for Epileptics. On eugenic fieldwork: Rafter, *White Trash*; Alice Wexler, *The Woman Who Walked into the Sea: Huntington's and the Making of a Genetic Disease* (New Haven: Yale University Press, 2008).

31. Charles Davenport, "Ecology of Epilepsy," *Archives of Neurological Psychiatry* 9 (1923): 554–66.

32. As cited in Davenport, "Ecology of Epilepsy," 557.

33. Davenport, "Ecology of Epilepsy," 557.

34. Davenport, "Ecology of Epilepsy," 557; Rafter, *White Trash*.

35. Douglas Baynton, "Defectives in the Land: Disability and American Immigration Policy, 1882–1924," *Journal of American Ethnic History* 24, no. 3 (2005): 31–44, 33.

36. H. D. Fabing and R. L. Barrow, *Epilepsy and the Law: A Proposal for Legal Reform in the Light of Medical Progress* (New York: Hoeber-Harper, 1956): 11–12, 28–29.

37. Paul Lombardo, *Three Generations, No Imbeciles: Eugenics, the Supreme Court, and Buck v. Bell* (Baltimore: Johns Hopkins University Press, 2008).

38. Horatio M. Pollock, "Epileptics in Institutions in the United States," *Psychiatric Quarterly* 1 (June 1927): 215–25. Ellen Dwyer has also written extensively about the Craig Colony: Ellen Dwyer, "The State and the Multiply Disadvantaged: The Case of Epilepsy," in *Mental Retardation in America: A Historical Reader,* ed. Steven Noll and James W. Trent (New York: New York University Press, 2004); Ellen Dwyer, "Stigma and Epilepsy," *Transactions and Studies of the College of Physicians of Philadelphia* 13, no. 4 (1991): 387–410; Dwyer, "Stories of Epilepsy."

39. Davenport and Weeks, "A First Study of Inheritance of Epilepsy."

40. For this widely cited statistic, based on 1930s census and World War I draftee data: William G. Lennox, *Epilepsy Explained* (Boston: Laymen's League Against Epilepsy, 1940).

41. See Karen Lystra's work based on the diaries and correspondence of Jean Clemens: Karen Lystra, *Dangerous Intimacy: The Untold Story of Mark Twain's Final Years* (Berkeley: University of California Press, 2004).

42. "Miss Jean Clemens Found Dead in Bath," *New York Times*, Dec. 25, 1909.

43. On gender, eugenics, and mothers: Kline, *Building a Better Race*; Molly Ladd-Taylor and Lauri Umansky, eds., *"Bad" Mothers: The Politics of Blame in Twentieth-Century America* (New York: New York University Press, 1998).

44. On the role of birth injuries, fevers, and childhood encephalitis in childhood seizures and epilepsy: Wilder Penfield, "Epilepsy and the Cerebral Lesions of Birth and Infancy," *Canadian Medical Association Journal* 41, no. 6 (1939): 527–34.

45. Hildred Ileene Jarvis, "A Study of the Detroit Experimental School for Epileptic Children" (MA diss., Institute of Public and Social Administration, University of Michigan, 1937), 9, 16.

46. For example: William T. Shanahan, "Convulsions in Infancy and Their Relationship, If Any, to Subsequent Epilepsy," *Psychiatric Quarterly* 2 (1928): 27–41, 39.

47. Pierre Berton, *The Dionne Years: A Thirties Melodrama* (New York: W.W. Norton, 1978).

48. "Emilie Dionne Victim of Epilepsy; Autopsy Reveals Family's Secret," *New York Times*, Aug. 8, 1954; "Death of a Dionne," *Life,* Aug. 1954, 35.

49. In interviews, surviving sisters later recounted being instructed not to tell doctors that Emilie had experienced seizures but only that she had "fainted." Jean-Yves Soucy with Cecile Dionne and Yvonne Dionne, *Family Secrets: The Controversial Story of the Dionne Quintuplets* (Toronto: Stoddart, 1997). On scrutiny of the Dionnes: Louise de Kiriline, *The Quintuplets' First Year: The Survival of the Famous Five Dionne Babies and Its Significance for All Mothers* (Toronto: Macmillan, 1936); *Protecting the Dionnes: A Valuable Manual for Mothers, with Useful Information about Scientific Modern Baby Care* (Toronto: Lysol Canada, 1936), Archives of Ontario, Toronto, ON; *The Country Doctor*, 20th Century Fox Films (1936).

50. Jefferson Lewis, *Something Hidden: A Biography of Wilder Penfield* (Toronto: Doubleday Canada, 1981), 118–25.

51. Fabing and Barrow, *Epilepsy and the Law*, 15–16.

52. Brand, *Dr. Kildare's Crisis*, 174.

53. Soucy et al., *Family Secrets*.

54. For example: A.G. patient record, 1936; M.S. patient record (patients anonymized), Montreal Neurological Institute (MNI), Montreal, QC.

55. Brand, *Dr. Kildare's Crisis*, 155.

56. For one of the earliest public pamphlets: Herbert Yahraes, *Epilepsy—The Ghost Is out of the Closet,* Public Affairs Pamphlet no. 98 (New York: Public Affairs Committee, 1944).

57. On experiences of polio and polio campaigns mid-century: Daniel Wilson, *Living with Polio: The Epidemic and Its Survivors* (Chicago: University of Chicago Press, 2005). Epilepsy differs from polio given that Dilantin, unlike the Salk vaccine (1955), predated advocacy efforts. Though epilepsy's campaigners sought to make epilepsy an issue of "public health," epilepsy was not infectious, and the polio campaign did not focus to the same degree on issues of stigma.

58. Throughout the 1920s, Harvard led in dietary and metabolic studies of epilepsy, followed by its work with electroencephalography. Anticonvulsant medications were developed at Columbia, and McGill led in new neurosurgeries for epilepsy. See Stanley Cobb and William G. Lennox, "The Work of the Harvard Epilepsy Commission," *Harvard Medical Alumni Bulletin* (June 1932), file E24, Countway Library of Medicine (CLM), Harvard University, Boston, MA.

59. Harvard Epilepsy Commission (HEC), Stanley Cobb Papers, 1934 and 1935 folders, CLM.

60. The application stated that epilepsy was not heritable in 85 percent of cases. See HEC 1935 folder, William G. Lennox Papers, CLM. The inklings of this change are also apparent in *Dr. Kildare's Crisis*, when Dr. Gillespie pulls Mary off the *Cabellero* and proclaims that epilepsy is "no more heritable than measles."

61. Frederic A. Gibbs and Erna L. Gibbs, *Atlas of Electroencephalography*, vol. 1 (Cambridge, MA: Addison-Wesley Press, 1950); William Lennox, "A Hereditary History of Epilepsy as Told by Relatives and Twins," *Journal of the American Medical Association* 146, no. 6 (1951): 529–36.

62. Parke, Davis and Company Records, National Museum of American History Archives Center (NMAH), Smithsonian Institution, Washington, DC; H. H. Merritt, and

T. J. Putnam, "Sodium Diphenyl Hydantoinate in the Treatment of Convulsive Disorders," *Journal of the American Medical Association* 111 (1938): 1068–73.

63. On medical optimism in the mid–twentieth century: Bert Hansen, *Picturing Medical Progress from Pasteur to Polio: A History of Mass Media Images and Popular Attitudes in America* (New Brunswick: Rutgers University Press, 2009); Allan M. Brandt and Martha Gardner, "The Golden Age of Medicine?" in *Companion to Medicine in the Twentieth Century,* ed. Roger Cooter and John Pickstone (New York: Routledge, 2003).

64. Eli Goldensohn et al., "The American Epilepsy Society: An Historic Perspective on 50 Years of Advances in Research," *Epilepsia* 38, no. 1 (2005): 124–150, 124.

65. The US Air Force used EEG to detect men susceptible to blackout, emotional breakdown, and "epileptoid tendencies." Kenton Kroker, "Washouts: Epilepsy and Emotions," in *Instrumental in War: Science, Research, and Instruments between Knowledge and the World*, ed. Steven A. Walton (Boston: Brill, 2005).

66. Wilder Penfield, "Neurosurgery in the War Period," in *Proceedings of the Second Annual Meeting, American-Soviet Medical Society*, Philadelphia, December 15, 1945.

67. The number of discharged servicemen with epilepsy was second only to the number of those discharged with schizophrenia. William Lennox, "Rehabilitation of Epileptic Service Men," *American Journal of Psychiatry* 100 (1943): 202–4. For rates of wounded servicemen who developed epilepsy: Herbert Yahraes, "Woman without Fear," *Woman's Home Companion*, February 1945, 20, 48.

68. William G. Lennox, *Epilepsy and Pearl Harbor* (Boston: Laymen's League Against Epilepsy, 1943). On the social dilemma of disabled veterans during and after the previous world war: Beth Linker, *War's Waste: Rehabilitation in World War I America* (Chicago: University of Chicago Press, 2011).

69. Physician's leagues for epilepsy included the National Association for the Study of Epilepsy and the Care and Treatment of Epileptics, organized in New York in 1898, and the American branch of the International League Against Epilepsy, formed in 1936, which would become the American Epilepsy Society in 1946. Goldensohn et al., "The American Epilepsy Society," 124.

70. Medical members of the LLAE included Stanley Cobb (neuropathology, Harvard Medical School); Temple Fay (neurology, Temple University); William Kerr (professor of medicine, University of California); Roger Lee (president of the American College of Physicians); Adolf Meyer (professor of psychiatry, Johns Hopkins University Medical School); Wilder Penfield (MNI); and Tracy Putnam (professor of neurology, Columbia University). See *Bulletin* no. 8 (January 1942), Laymen's League Against Epilepsy, box 3, folder 9, William G. Lennox Papers, CLM.

71. The statistic given in letters between LLAE head, Dr. Lennox, and its president was that roughly four-fifths of members were patients. Letters from William G. Lennox to Francis B. Riggs, Dec. 17 and 24, 1940, box 3, folder 8, LLAE 1938–1941, William G. Lennox Papers, CLM. On membership, for which fees ranged between $1 and $5 a year: Lennox, *Epilepsy Explained*.

72. Letter from Lennox to members of the LLAE stating "names of members will not be made public," Aug. 4, 1939, HEC 1939 folder, William G. Lennox Papers, CLM. On anonymity of members, see also *Bulletin* no. 8 (Jan. 1942), Laymen's League Against

Epilepsy, William G. Lennox Papers, CLM. On the Roosevelts and epilepsy: Steven Lomazow and Eric Fettmann, *FDR's Deadly Secret* (New York: Perseus, 2009).

73. Letter from Francis B. Riggs to prospective doctor members of the LLAE, n.d., box 3, folder 9, LLAE 1938–1941, William G. Lennox Papers, CLM.

74. Brooks Potter reported a rate of five to six letters a day. Yahraes, "Woman without Fear," 20.

75. Yahraes, "Woman without Fear," 20."

76. Letter from Francis B. Riggs to LLAE members, Jan. 1941, Laymen's League Against Epilepsy, box 3, folder 9, William G. Lennox Papers, CLM.

77. *Bulletin* no. 8 (Jan. 1942), Laymen's League Against Epilepsy, William G. Lennox Papers, CLM.

78. Lennox, *Epilepsy Explained*.

79. Royalties from the book went to the Laymen's League. *Bulletin* no. 6 (Jan. 1941), Laymen's League Against Epilepsy, William G. Lennox Papers, CLM.

80. *Bulletin* no. 8 (Jan. 1942), Laymen's League Against Epilepsy, William G. Lennox Papers, CLM.

81. *Bulletin* no. 8 (Jan. 1942), Laymen's League Against Epilepsy, William G. Lennox Papers, CLM.

82. *Bulletin* no. 8 (Jan. 1942), Laymen's League Against Epilepsy, William G. Lennox Papers, CLM.

83. *Bulletin* non. 8 (Jan. 1942), Laymen's League Against Epilepsy, William G. Lennox Papers, CLM.

84. *Bulletin* no. 8 (Jan. 1942), Laymen's League Against Epilepsy, William G. Lennox Papers, CLM.

85. Spencer, "No Wonder Epileptics Are Bitter."

86. The term "near normal" emerged to publicly describe people with epilepsy in the early postwar period. Disability scholars have written extensively about the problem of "normalcy." For instance: Lennard J. Davis, *Enforcing Normalcy: Disability, Deafness, and the Body* (London: Verso, 1995).

87. On the American Epilepsy Society: Goldensohn et al., "The American Epilepsy Society." For an example of the kind of outreach and fundraising present by the 1950s: "Epilepsy Unit Gains by Celebrity Ball," *New York Times*, Jan. 1, 1955, 17.

88. In 1942, Francis Briggs of the Laymen's League Against Epilepsy first advocated for coverage of epilepsy in popular magazines: *Bulletin* no. 8 (Jan. 1942), Laymen's League Against Epilepsy, William G. Lennox Papers, CLM.

89. Yahraes, *Epilepsy—The Ghost Is out of the Closet*.

90. For example: "Epilepsy Isn't a Curse!" *Better Homes and Gardens*, Mar. 1948, 180–85. See also William Lennox, "Epilepsy: The Hopeful Disorder," *Parent's Magazine*, Feb. 1946, 32, 124–27; Morris Fishbein, "Nervous and Convulsive Disorders in Children," *American Home,* Dec. 1946, 77–80; "Some Plain English on Epilepsy," *Science Digest* 17 (1945): 82–83; "To Aid Epilepsy Group—Robert Montgomery Joins Drive to End 'Hush-Hush' Stigma," *New York Times*, Oct. 14, 1952, 33; Robert Bassett, "The Problem of Epilepsy," *Hygeia* 24 (1946): 520–21.

91. Milton Silverman, "We Can Lick Epilepsy," *Saturday Evening Post*, Jan. 17, 1948, 107–14.

92. Matthias M. Krenn, "Social Outcast—Age 9," *Today's Health* 29, no. 8 (1951): 28–30, 28.

93. William G. Lennox, "Epilepsy—A Problem of Public Health," *American Journal of Public Health* 41 (May 1951): 533–89.

94. William G. Lennox, *Epilepsy and Related Disorders* (Boston: Little, Brown, 1960), 974.

95. For a source that examines race, class, ethnicity, and religion in exploring epilepsy: Horace B. Eldred, "Epilepsy in Adult Males: Impact on Wives" (MA thesis, Faculty of the School of Social Work, University of Southern California, 1959), 12, 17, 37–39.

96. For a summary of these exclusions in the popular press: Gerald Walker, "Epilepsy: Medical Triumph, Social Tragedy," *Pageant* 16, no. 6 (Dec. 1960), 93–96.

97. As reported of a 1942 Lennox and Cobb study, which collected employment information on 1,105 people with epilepsy, in Lennox, *Epilepsy and Related Disorders,* 952.

98. Lennox, "Rehabilitation of Epileptic Service Men," 203; Lennox, *Epilepsy and Related Disorders*, 997–998. A number of historians have examined exclusions of the 1944 G.I. Bill along lines of race and sexuality. For example: Margot Canaday, "Building a Straight State: Sexuality and Social Citizenship under the 1944 G.I. Bill," *Journal of American History* 90, no. 3 (2003): 935–57.

99. On the comparable League of the Physically Handicapped during World War II: Paul Longmore, "Making Disability an Essential Part of American History," *OAH Magazine of History* 23, no. 3 (2009): 11–15.

100. David S. Schechter, "Rehabilitation of Epileptics," *Journal of Rehabilitation* 13, no. 3 (1947): 28–29. Given the dangers and potential for accidents associated with many jobs, such groups gradually advocated for revisions to liability laws that disincentivized employers from hiring people with epilepsy.

101. Yahraes, *Epilepsy—The Ghost Is out of the Closet*, 24–25.

102. Fabing and Barrow, *Epilepsy and the Law*, 35–61.

103. Fabing and Barrow, *Epilepsy and the Law*; Lennox, *Epilepsy and Related Disorders*, 974.

104. Eldred, "Epilepsy in Adult Males"; Virginia-May Schacht, "Epilepsy in Adult Males: Impact on Wives" (MA thesis, Faculty of the School of Social Work, University of Southern California, 1959), 49–51.

105. Eldred, "Epilepsy in Adult Males," 16, 38.

106. Schacht, "Epilepsy in Adult Males," 49.

107. Eldred, "Epilepsy in Adult Males," 23–24, 26–27.

108. Eldred, 17, 37.

109. Eldred, 28–31, 35–36.

110. On a similar type of popular postwar coverage featuring celebrity patients: Barron Lerner, *When Illness Goes Public: Celebrity Patients and How We Look at Medicine* (Baltimore: Johns Hopkins University Press, 2006).

111. Discussion of famous people with epilepsy emerged in the 1940s and made appearances in approximately half of the popular postwar articles surveyed by the author.

112. Such stories more commonly featured young men and questions of work. Female characters tended to be wives or school-age girls.

113. Edward Dengrove and Doris Kulman, "Should Epileptics Marry?" *Today's Health*, Sept. 1955, 19, 54–55.

114. Krenn, "Social Outcast—Age 9."

115. Herb Bailey, "The Truth about Epilepsy," *Science Digest*, Sept. 1950, 73–76; M. G. Peterman, "Convulsions in Childhood," *Today's Health*, July 1956, 34–37.

116. Silverman, "We Can Lick Epilepsy," 108.

117. On postwar normalcy: Sarah Igo, *The Averaged American: Surveys, Citizens, and the Making of a Mass Public* (Cambridge: Harvard University Press, 2007); Anna Creadick, *Perfectly Average: The Pursuit of Normality in Postwar America* (Amherst: University of Massachusetts Press, 2010); Mary Louise Adams, *The Trouble with Normal: Postwar Youth and the Making of Heterosexuality* (Toronto: University of Toronto Press, 1997).

118. "Ultimate Hope of Ending Epilepsy Put Shifting Ions in Tiny Gene," *New York Times*, May 10, 1937. On the emergence of intelligence testing and its place in eugenics: Leila Zenderland, *Measuring Minds: Henry Herbert Goddard and the Origins of American Intelligence* (Cambridge: Cambridge University Press, 1998).

119. Letter from Lennox to Talbot, May 21, 1937, HEC folder 1937, William G. Lennox Papers, CLM.

120. Yahraes, "Woman without Fear," 20.

121. Yahraes, "Woman without Fear," 20."

122. For example: "Epilepsy Isn't a Curse!" *Better Homes and Gardens*, 180.

123. Frederic Gibbs and Frederick Stamps, *Epilepsy Handbook* (Springfield, IL: Charles C. Thomas, 1958), 3–4.

124. Dr. Lennox, for instance, described epilepsy in an early pamphlet as "no more than a disturbance of electrical pulsations of the brain"; Lennox, *Epilepsy Explained*.

125. As quoted under the heading "If Epilepsy Has a Bad Name, Why Not Change It?" in Yahraes, *Epilepsy—The Ghost Is out of the Closet*, 5.

126. Historians of sexuality have written at length about the double, secret, and closeted lives of gay men and women. For a key example focused on this era, see Berube, *Coming Out under Fire*. On disability passing, see also Brune and Wilson, *Disability Passing*.

127. Tracy J. Putnam, *Convulsive Seizures: How to Deal with Them* (Philadelphia: Lippincott, 1943), 24.

128. For instance: Peter Wolf, "Epilepsy in Literature," *Epilepsia* 36, no. S1 (1995): S12–S17, S12; Jeannette Stirling, *Representing Epilepsy: Myth and Matter* (Liverpool: Liverpool University Press, 2010).

129. Yahraes, "Woman without Fear," 20.

130. Lennox, *Epilepsy Explained*.

131. For example, one magazine article describing a man with epilepsy stated, "He is a part-time-handicapped person. His handicap lasts about five to eight minutes once or twice a month"; "Epilepsy Isn't a Curse!" *Better Homes and Gardens*, 180.

132. Herbert and Dixie Yahraes, "You'd Never Know Our Daughter Is an Epileptic," *Collier's*, Nov. 26, 1949, 18–19.

133. Putnam, *Convulsive Seizures,* 63–64.

134. William T. Shanahan, "The Problem of Epilepsy in New York State," *Psychiatric Quarterly* 1, no. 2 (1927): 160–83.

135. Putnam, *Convulsive Seizures,* 70–71.

136. As cited and discussed in Lennox, *Epilepsy and Related Disorders*, 1060.

137. For this Dilantin advertisement, as it appeared in one medical journal: "Symbols of Significance: Dilantin Sodium," 1944–1949, *American Journal of the Medical Sciences*, Health Advertisements Database from Ebling Sources (HADES), Ebling Library for the Health Sciences Collection, University of Wisconsin, Madison, https://search.library.wisc.edu/digital/AQS4EQKY7IWWWF8R.

138. "Chance for Elizabeth," *Time*, Nov. 29, 1948, 73.

139. Letter F to Dr. Wilder Penfield, Dec. 1948, box 44, C/G 7, Osler Library for the History of Medicine (OLHM), McGill University, Montreal, QC.

140. Letter C to Dr. Wilder Penfield, Apr. 1948, box 44, C/G 7, Wilder Penfield Fonds, OLHM.

141. Letter H to Dr. Wilder Penfield, Dec. 1948, box 44, C/G 7, Wilder Penfield Fonds, OLHM.

142. For example, the records of Wilder Penfield's patients reveal people in a variety of jobs and social roles, both before and after their surgeries; patient records, Wilder Penfield Fonds, OLHM.

143. Putnam, *Convulsive Seizures,* 152.

144. *Annual Report of the White Special School, 1949–50,* 90, Detroit Public School Records, Acc #681, Walter P. Reuther Library (WPR), Wayne State University, Detroit.

145. For instance: *Annual Report of the White Special School, 1949–1950,* 90; Marshall Shields, "Epileptic Drivers," *Chicago Daily Tribune*, Nov. 30, 1957, 12.

146. For a recent example and discussion of invisible illness and disability, including the kinds of conditions and disabilities covered by the concept: O'Rourke, *The Invisible Kingdom.*

Chapter 2. Technology and Transparency

1. "Geniuses Aid Tests of Brain Processes," *New York Times*, Feb. 24, 1951, 9.

2. For postwar characterizations of "epileptic geniuses": Bryant J. Ernest, *Genius and Epilepsy: Brief Sketches of Great Men Who Had Both* (Concord, MA: Ye Old Depot Press, 1953).

3. The idea that epilepsy legitimized and launched EEG has been argued by Cornelius Borck: Christoph Gradmann, "Review of *Hirnstrome: Eine Kulturgeschichte der Elektro-enzephalographie*, by Cornelius Borck," *Isis* 98, no. 4 (2007): 843–44.

4. Steven M. Spencer, "No Wonder Epileptics Are Bitter," *Saturday Evening Post*, Mar. 28, 1953, 154.

5. On the social construction of technology: Wiebe E. Bijker et al., eds., *The Social Construction of Technological Systems: New Directions in the Sociology and History of Technology* (Cambridge: MIT Press, 1987).

6. William G. Lennox, "New Light on Ancient Ills," chap. 1 in *Science and Seizures: New Light on Epilepsy and Migraine* (New York: Paul B. Hoeber, 1941), 8.

7. On epilepsy and the study of the mind as "less organic": Kenton Kroker, *The Sleep of Others and the Transformations of Sleep Research* (Toronto: University of Toronto Press, 2007), 262.

8. On asymptomatic "carriers" of infectious disease: Judith Walzer Leavitt, *Typhoid Mary: Captive to the Public's Health* (Boston: Beacon Press, 1996).

9. On the "masked" epileptic: Wolfgang Saxon, "Frederic A. Gibbs, 89, Neurologist and Top Researcher on Epilepsy," *New York Times*, Oct. 23, 1992. On disability "passing": Jeffrey A. Brune and Daniel J. Wilson, eds., *Disability and Passing: Blurring the Lines of Identity* (Philadelphia: Temple University Press, 2013), 1–12.

10. Keith Wailoo, *Drawing Blood: Technology and Disease Identity in Twentieth-Century America* (Baltimore: Johns Hopkins University Press, 1997); Robert Aronowitz, *Unnatural History: Breast Cancer and American Society* (New York: Cambridge University Press, 2007); Joel D. Howell, *Technology in the Hospital: Transforming Patient Care in the Early Twentieth Century* (Baltimore: Johns Hopkins University Press, 1995); Ian Hacking, "Making People Up," in *Science Studies Reader,* ed. Mario Biagioli (New York: Routledge, 1999).

11. On technologies of detection and their social effects: Ken Alder, *The Lie Detectors: History of an American Obsession* (New York: Free Press, 2007); Leavitt, *Typhoid Mary*.

12. Kenton Kroker argues that EEG helped frame sleep as a process regulated by the brain and as an object of laboratory analysis. Andrew Pickering suggests that EEG became a tool for measuring the cybernetic brain itself through the work of British physiologist and robotician Walter Grey. Cornelius Borck's German-language work on Hans Berger studies the early development of the technology. Kroker, *Sleep of Others*; Andrew Pickering, *The Cybernetic Brain: Sketches of Another Future* (Chicago: University of Chicago Press, 2010); Cornelius Borck, *Hirnstrome: Eine Kulturgechichte der Elektroenzephalographie* (Gottingen: Wallstein, 2005).

13. On postwar normalcy: Anna G. Creadick, *Perfectly Average: The Pursuit of Normality in Postwar America* (Amherst: University of Massachusetts Press, 2010); Sarah Igo, *The Averaged American: Surveys, Citizens, and the Making of a Mass Public* (Cambridge: Harvard University Press, 2007); David Serlin, *Replaceable You: Engineering the Body in Postwar America* (Chicago: University of Chicago Press, 2004).

14. Lennox, "New Light on Ancient Ills," 8.

15. Frederic A. Gibbs and Erna L. Gibbs, *Atlas of Electroencephalography*, vol. 1 (Cambridge, MA: Addison-Wesley Press, 1950), 18.

16. Gibbs and Gibbs, *Atlas*, vol. 1, 18.

17. Gibbs and Gibbs, *Atlas of Electroencephalography*, vol. 3 (Cambridge, MA: Addison-Wesley, 1964), 1.

18. Letter to HEC members from William Lennox, Jan. 1934, Harvard Epilepsy Commission Papers, Layman's League Against Epilepsy, Correspondence, E24.10, Countway Library of Medicine (CLM), Center for the History of Medicine, Harvard University, Boston, MA.

19. F. A. Gibbs et al., "The Electroencelphalogram in Epilepsy and in Conditions of Impaired Consciousness," *Archives of Neurology and Psychiatry* 34 (1935): 1133, and as

recounted in Stanley Cobb, "Obituary: William Gordon Lennox, 1884–1960," *Archives of Neurology* (1960): 123–25, 124.

20. "Hans Berger; Discoverer of Electrical 'Brain Waves' in Humans Was 68," *New York Times*, June 10, 1941.

21. In the 1870s, Gustav Fritsch and Eduard Hitzig observed the motor excitability of the cortex in living animals. This coincided with what historians generally identify as the rise of modern neurology. It is also the period in which epilepsy came to be theorized as a condition of electrical activity in the brain. Gibbs and Gibbs, *Atlas of Electroencephalography*, vol. 1, 1; Owsei Temkin, *The Falling Sickness: A History of Epilepsy from the Greeks to the Beginnings of Modern Neurology* (Baltimore: Johns Hopkins University Press, 1945), 288.

22. "Hans Berger," *New York Times*; Lennox, *Science and Seizures*, 86.

23. Gibbs and Gibbs, *Atlas*, vol. 1, dedication page.

24. On the history of EEG and its role in experimental interwar psychology and neuropsychiatry: David E. Millet, "Wiring the Brain: From the Excitable Cortex to the EEG, 1870–1940" (PhD diss., University of Chicago, 2001); Kroker, *Sleep of Others*, 264–65; W. T. Blume, "History of Electroencephalography as a Pre-Surgical Evaluation Tool: The Pre-Berger Years," in *Textbook of Epilepsy Surgery*, ed. Hans O. Luders (New York: Taylor and Francis, 2008) 174–76, 174.

25. Gibbs and Gibbs, *Atlas*, vol. 1, 1.

26. Willam Lennox, "Epilepsy and the Epileptic," *JAMA* 162, no. 2 (Sept. 8, 1956): 118–19, 118.

27. "The World of the Harvard Epilepsy Commission," *Harvard Medical Alumni Bulletin*, June 1932, 55, Harvard Epilepsy Commission Papers, CLM.

28. Letter of the Harvard Epilepsy Commission, June 19, 1935, Harvard Epilepsy Commission Papers, CLM.

29. Steven J. Zottoli, "The Origins of The Grass Foundation," *Biology Bulletin* 201 (Oct. 2001): 218–26.

30. Letter of the Harvard Epilepsy Commission, Apr. 13, 1936, Harvard Epilepsy Commission Papers, CLM.

31. Letter of the Harvard Epilepsy Commission, Apr. 6, 1935, Harvard Epilepsy Commission Papers, CLM.

32. Letter of the Harvard Epilepsy Commission, Apr. 6, 1935; on the brain and collecting: Warwick Anderson, *The Collectors of Lost Souls: Turning Kuru Scientists into Whitemen* (Baltimore: Johns Hopkins University Press, 2008); Alison Winter, *Memory: Fragments of a Modern History* (Chicago: University of Chicago Press, 2012).

33. Lennox, *Science and Seizures*, 88.

34. "The World of the Harvard Epilepsy Commission," 55.

35. Zottoli, "Origins of The Grass Foundation," 219.

36. On early EEG research on the effects of various drugs: F. A. Gibbs et al., "Effect on the Electroencelphalogram of Certain Drugs Which Influence Nervous Activity," *Archives of Internal Medicine* 60, no. 1 (1937): 154–66.

37. On the use of EEG in developing anticonvulsants: Suzanne Corkin, *Permanent Present Tense: The Man with No Memory and What He Taught the World* (New York: Penguin, 2013), 10. On the development of Dilantin and new research methods: Lewis P. Rowland,

The Legacy of Tracy J. Putnam and H. Houston Merritt: Modern Neurolgoy in the United States (New York: Oxford University Press, 2009), 11.

38. On the increased use of EEG in neurosurgery as a "window" into the skull: "More Patients Cared for by Institute: Neurological Work Taxes Capacity of Hospital," *Montreal Daily Star,* Sept. 30, 1939, 3–4, 3.

39. Herbert Jasper's early years commuting to and from the Montreal Neurological Institute for weeks at a time to examine Wilder Penfield's surgical epilepsy patients is detailed in some letters between Penfield and patients. Letter from Wilder Penfield to J.V.'s father, Mar. 9, 1938, J.V. patient record, 1936, Montreal Neurological Institute-Hospital (MNI), Montreal, QC.

40. Jefferson Lewis, *Something Hidden: A Biography of Wilder Penfield* (Toronto: Doubleday, 1981), 193–96.

41. Blume, "History of Electroencephalography," 174–75.

42. Herbert Jasper, "Charting the Sea of Brain Waves," *Journal of Clinical Neurophysiology* 14 (1997): 464–69; Herbert H. Jasper, "History of the Early Development of Electroencephalography and Clinical Neurophysiology at the Montreal Neurological Institute: The First 25 Years, 1939–1964," *Canadian Journal of Neurological Sciences* 18, no. 4 (1991): 533–53, 533–36.

43. On the early popular responses to EEG in Germany and the United States, respectively: Borck, *Hirnstrome*; Kroker, *Sleep of Others,* 284–85.

44. "Man's Mind Traced by His Electricity: Radio-Active Index of Brain and Nerves Described by Dr. Hallowell Davis: Also Tests Identity," *New York Times*, Apr. 23, 1937.

45. On electricity and the body in popular culture: Carolyn Thomas de la Pena, *The Body Electric: How Strange Machines Built the Modern American* (New York: New York University Press, 2003); on neurology and the modern self: Fernando Vidal, "Brainhood, Anthropological Figure of Modernity," *History of the Human Sciences* 22, no. 1 (2009): 5–36; Laura Salisbury and Andrew Shail, *Neurology and Modernity: A Cultural History of Nervous Systems, 1800–1950* (London: Palgrave MacMillan, 2010).

46. "Man's Mind Traced by His Electricity."

47. Letter from Dr. Lennox to Dr. Talbot, May 21, 1937, Harvard Epilepsy Commission Papers, CLM.

48. Herbert Yahraes, *Epilepsy—The Ghost Is out of the Closet*, Public Affairs Pamphlet no. 98 (New York: Public Affairs Committee, 1944), 10.

49. Yahraes, *Epilepsy—The Ghost Is out of the Closet*, 10.

50. Yahraes, *Epilepsy—The Ghost Is out of the Closet*, 10.

51. *Annual Report of the White Special School, 1944*, 8, Detroit Public School Records, Acc #681, Walter P. Reuther Library (WPR), Archives of Labor and Urban Affairs, Wayne State University, Detroit, MI.

52. Letter from Alice Metzner to Dr. William Lennox, Oct. 28, 1941, B MS c 113, box 3, folder 9, Harvard Epilepsy Commission Papers, CLM.

53. Letters from Alice Metzner to Dr. Lennox, July 16, 1941, and Oct. 28, 1941, Harvard Epilepsy Commission Papers, CLM.

54. The materials needed to build EEG machines were placed on A-1 priority by the US government during World War II, meaning that they were urgently important for

defense. In spite of this, the school managed to secure an EEG machine by 1943. Letter from Alice Metzner to Dr. Lennox, Oct. 28, 1941, Harvard Epilepsy Commission Papers, CLM.

55. *Annual Report of the White Special School, 1945–1946*, 3, WPR.

56. *Annual Report of the White Special School, 1945–1946*, 3, WPR.

57. These were typically private patients from Boston's outpatient hospitals between 1929 and 1944. Lennox, *Science and Seizures*, 87; Gibbs and Gibbs, *Atlas*, vol. 1, 106–16.

58. Interestingly, "normal" was derived from "abnormal." Historians in other contexts have identified this as a primary feature of modern neurology: Stephen Jacyna, *Lost Words: Narratives of Language and the Brain, 1825–1926* (Princeton: Princeton University Press, 2000); Stephen Jacyna and Stephen Casper, *The Neurological Patient in History* (Rochester: University of Rochester Press, 2012).

59. Lennox, *Science and Seizures*, 87.

60. Virginia Evans, *The Lovely Season* (New York: Appleton-Century-Crofts, 1952), 111.

61. Evans, *The Lovely Season,* 128.

62. Evans, *The Lovely Season,* 111.

63. As Lennox recounted to a colleague, "Epileptic seizures, whether there is muscle movement or not, are accompanied by tremendous alterations in the normal electrical potential of the brain," and moreover, "in some cases abnormal changes are present even when the patient feels or shows no evidence of having a seizure." Letter from William G. Lennox, Apr. 17, 1936, Letters of the Harvard Epilepsy Commission, Harvard Epilepsy Commission Papers, CLM; Lennox et al., "The Inheritance of Epilepsy as Revealed by the Electroencephalograph," *JAMA* 113, no. 11 (Sept. 1939): 9.

64. Letter from Dr. Davis to Dr. Cobb, June 19, 1935, Letters of the Harvard Epilepsy Commission, Harvard Epilepsy Commission Papers, CLM.

65. Lennox, *Science and Seizures*, 90.

66. Lennox, *Science and Seizures*, 88, 90.

67. Lennox, *Science and Seizures*, 88.

68. Frederic A. Gibbs and Frederick W. Stamps, *Epilepsy Handbook* (Springfield, IL: Charles Thomas, 1958), 3.

69. "We Can Lick Epilepsy," *Saturday Evening Post*, Jan. 17, 1946, 109.

70. Lennox, "Epilepsy and the Epileptic," 118.

71. Lennox, *Science and Seizures*, 86–93; Lennox, *Epilepsy and Related Disorders* (Boston: Little, Brown, 1960), 914.

72. For example, see Robert Bassett, "The Problem of Epilepsy," *Hygeia* 24 (1946): 520–21; William Lennox, "Epilepsy: The Hopeful Disorder," *Parents' Magazine*, Jan. 1946, 124–27; "Some Plain English on Epilepsy," *Science Digest* 17 (1945): 82–83. On media and medicine more generally in this period: Bert Hansen, *Picturing Medical Progress from Pasteur to Polio: A History of Mass Media Images and Popular Attitudes in America* (New Brunswick: Rutgers University Press, 2009); Barron Lerner, *When Illness Goes Public: Celebrity Patients and How We Look at Medicine* (Baltimore: Johns Hopkins University Press, 2006).

73. "Some Plain English on Epilepsy," *Science Digest*, 82.

74. Letter from Dr. Davis to Dr. Cobb, June 19, 1935, Letters of the Harvard Epilepsy

Commission, Harvard Epilepsy Commission Papers, CLM; "Three New Drugs to Aid War on Epilepsy," *Boston Post*, June 5, 1937.

75. For example: "We Can Lick Epilepsy," 22.

76. "We Can Lick Epilepsy," 22.

77. Matthias M. Krenn, "Social Outcast—Age 9," *Today's Health* 29 no. 8 (1951): 28–30, 29.

78. Howell, *Technology in the Hospital*, 1–10.

79. Dorothy Berger's seizures were thought to be hysterical because her explanation of having them after looking through window screens seemed implausible to her doctors. Letter from Dr. Lennox to Dr. Talbot, Feb. 26, 1938, Letters of the Harvard Epilepsy Commission, Harvard Epilepsy Commission Papers, CLM.

80. *Annual Report of the White Special School, 1947–1948*, 103; *Annual Report of the White Special School, 1950–1951*, 99, WPR.

81. *Annual Report of the White Special School, 1950–1951*, 99, WPR.

82. Lennox, *Science and Seizures*, 90.

83. Lennox, *Science and Seizures*.

84. Lennox, *Epilepsy and Related Disorders*, 974.

85. In 1937, Lennox and the Gibbses made their first recordings of the relatives of patients with epilepsy at Boston Children's Hospital and found that 60 percent had a "definite" dysrhythmia. They also found that 12 percent of the population, amounting to almost 15 million Americans, had a predisposition to epilepsy or an "allied disorder." W. G. Lennox et al., "Inheritance of Cerebral Dysrhythmia and Epilepsy," *Archives of Neurology and Psychiatry* 44, no. 6 (Dec. 1940): 1155–83, 1181.

86. Lennox, *Science and Seizures*, 90.

87. See "What Is Normal?" chap. 10 in Gibbs and Gibbs, *Atlas*, vol. 1.

88. A study of 1,359 clinical cases by 1935 led Lennox and the Gibbses to conclude that "epilepsy is not a hereditary disease in 85% of cases." For a discussion of this research: Application Letter to the Milbank Foundation, Letter of the Harvard Epilepsy Commission, June 19, 1935, Harvard Epilepsy Commission Papers, CLM.

89. Lennox et al., "The Inheritance of Epilepsy as Revealed by the Electroencephalograph," 9.

90. Lennox, *Science and Seizures*, 87.

91. Lennox et al., "Inheritance of Cerebral Dysrhythmia," 1178–79.

92. Lennox et al., "The Inheritance of Epilepsy as Revealed by the Electroencephalograph," 9. On epilepsy as a "family disease": Temkin, *The Falling Sickness*; Karen Lystra, *Dangerous Intimacy: The Untold Story of Mark Twain's Final Years* (Berkeley: University of California Press, 2004); Ellen Dwyer, "Stories of Epilepsy, 1880–1930," in *Framing Disease: Studies in Cultural History*, ed. Charles Rosenberg and Janet Golden (New Brunswick: Rutgers University Press, 1992).

93. Lennox, *Science and Seizures*, 92.

94. According to Lennox, as much as 1 percent of the population had epilepsy, many more were unknown carriers, and fewer than 10 percent of all people with epilepsy lived in institutions. William G. Lennox, *Epilepsy Explained* (Boston: Laymen's League Against Epilepsy, 1940); "Triumph of Epilepsy," *New York Times*, June 3, 1940.

95. "Triumph of Epilepsy," *New York Times.*

96. "Triumph of Epilepsy," *New York Times.*

97. Saxon, "Frederic A. Gibbs, 89."

98. Hans Hoff and Helen Clarke, *Will My Child Have Epilepsy?* (New York: National Association to Control Epilepsy, 1949), 1, New York Academy of Medicine Archives and Manuscripts (NYAM), New York.

99. Hoff and Clarke, *Will My Child Have Epilepsy?*

100. Gerald Walker, "Epilepsy—Medical Triumph, Social Tragedy," *Pageant,* Dec. 1960, 91–97, 91.

101. Spencer, "No Wonder Epileptics Are Bitter," 154; Lennox, *Epilepsy Explained.*

102. On the multiple "preventative" measures of EEG: Lennox, *Epilepsy and Related Disorders,* 1005. On the search for "hidden" cases of "cerebral dysrhythmia": Kroker, *Sleep of Others,* 285.

103. On deliberations regarding the use of EEG in driver's licensing: Dr. E. Schwade to Mr. Charles Crownhart, Jan. 27, 1949, State Medical Society Records, box 88, folder 6, "Epilepsy, 1947–1956," Wisconsin Historical Society Library (WHSL), University of Wisconsin, Madison, WI. On the exclusion of immigrants with epilepsy starting in 1903: Douglas C. Baynton, "Defectives in the Land: Disability and American Immigration Policy, 1882–1924," *Journal of American Ethnic History* 24, no. 3 (2005): 31–44, 33.

104. Jasper, "History of the Early Development of Electroencephalography," 536–37.

105. Lennox, "Epilepsy: The Hopeful Disorder." On disability screening during World War I more generally: Beth Linker, "Feet for Fighting: Locating Disability and Social Medicine in First World War America," *Social History of Medicine* 20, no. 1 (2007): 90–109.

106. William G. Lennox, *Epilepsy and Pearl Harbor* (Boston: Laymen's League Against Epilepsy, 1943).

107. William G. Lennox, "Rehabilitation of Epileptic Service Men," *American Journal of Psychiatry* 100 (Sept. 1943): 202–4.

108. Lennox, "Rehabilitation of Epileptic Service Men."

109. Letter from Dr. William Lennox to Alice Metzner, Oct. 29, 1941, Harvard Epilepsy Commission Papers, CLM.

110. Kenton Kroker, "Washouts: Epilepsy and Emotions," in *Instrumental in War: Science, Research, and Instruments Between Knowledge and the World,* ed. Steven A. Walton (Boston: Brill, 2005), 302–3.

111. On "seizures as predominantly a disorder of youth": Lennox, "Epilepsy—A Problem of Public Health," *American Journal of Public Health* 41 (May 1951): 533–89, 533.

112. *Annual Report of the White Special School, 1945–1946,* 3, WPR.

113. Letter from Alice Metzner to Dr. William Lennox, Oct. 28, 1941, Harvard Epilepsy Commission Papers, CLM; *Annual Report of the White Special School, 1945–1946,* 3, WPR.

114. *Annual Report of the White Special School, 1945–1946,* 4, 9, WPR.

115. On the 1943 case of Edwin Codarre: William I. Garfinkel, "The Case of Edwin Codarre: An Overview," unpublished manuscript, shared with author.

116. W. L. Neustatter, "Elwell the Epileptic," in *The Mind of the Murderer* (New York: Philosophical Library, 1957).

117. Alder calls the polygraph "America's mechanical consciousness," reflecting desires to "fashion society by non-coercive mechanisms": Alder, *The Lie Detectors*, xiv.

118. "Triumph of Epilepsy," *New York Times*, 12.

119. On eugenics and medical marriage counseling more generally: Wendy Kline, *Building a Better Race: Gender, Sexuality, and Eugenics from the Turn of the Century to the Baby Boom* (Berkeley: University of California Press, 2005); Alexandra Stern, *Eugenic Nation: Faults and Frontiers of Better Breeding in Modern America* (Los Angeles: University of California Press, 2005).

120. Lennox et al., "Inheritance of Cerebral Dysrhythmia," 1183; Waldemar Kaempffert, "Studies of Epilepsy: Review of *Science and Seizures*, by William G. Lennox," *New York Times*, Mar. 23, 1941, BR29.

121. Lennox et al., "Inheritance of Cerebral Dysrhythmia," 1183.

122. Evans, *The Lovely Season*, 112.

123. Evans, *The Lovely Season*, 143.

124. Evans, *The Lovely Season*, 143.

125. Evans, *The Lovely Season*, 143.

126. On the neurological extrapolation of "normal" from "abnormal": Jacyna, *Lost Words*.

127. "Delinquents' Brains: Electrical Waves Are Different from Those of Normal Youth," *New York Times*, Oct. 17, 1954, E9.

128. "Delinquents' Brains," *New York Times*.

129. "Delinquents' Brains," *New York Times*.

130. Lennox, *Science and Seizures*, 87.

131. "Delinquents' Brains," *New York Times*.

132. *Annual Report of the White Special School, 1941*, 4, WPR.

133. *Annual Report of the White Special School, 1947–1948*, 99, 100, WPR; *Annual Report of the White Special School, 1945–1946*, 8, WPR.

134. *Annual Report of the White Special School, 1943–1944*, 2–3, WPR; *Annual Report of the White Special School, 1952–1953*, 3, WPR.

135. Letter from Alice Metzner to Dr. Lennox, Oct. 28, 1941, Harvard Epilepsy Commission Papers, CLM; Dr. Lennox to Alice Metzner, Oct. 29, 1941, Harvard Epilepsy Commission Papers, CLM.

136. Letter from Dr. Lennox to Alice Metzner, Oct. 29, 1941.

137. Letter from Dr. Lennox to Alice Metzner, Oct. 29, 1941.

138. Letter from Dr. Lennox to Alice Metzner, Oct. 29, 1941.

139. Lennox et al., "Inheritance of Cerebral Dysrhythmia," 1169.

140. Kroker, "Washouts," 316.

141. Lennox, *Science and Seizures*, 92.

142. Kroker, "Washouts," 321; Garfinkel, "The Case of Edwin Codarre."

143. Alfred C. Kinsey et al., *Sexual Behavior in the Human Female* (Bloomington: Indiana University Press, 1953), 627–713.

144. Letter from Dr. Abraham Mosovich to Dr. Alfred Kinsey, Mar. 9, 1954, Kinsey Correspondence Collection, box 13, Abraham Mosovich folder, Kinsey Institute Library &

Special Collections (KIL), Bloomington, IN; Kinsey et al., *Sexual Behavior in the Human Female*, 704, 706–7.

145. On female orgasm and therapeutics in the nineteenth century: Rachel P. Maines, *The Technology of Orgasm: "Hysteria," the Vibrator, and Women's Sexual Satisfaction* (Baltimore: Johns Hopkins University Press, 1999).

146. First publication in English: Abraham Mosovich and Alberto Tallaferro, "Studies on EEG and Sex Function Orgasm," *Diseases of the Nervous System* (July 1954): 218–20. The study consisted of EEG records taken during masturbation in three male and three female subjects, in whom "careful screening" was made for no sign of a personal or familial history of epilepsy.

147. Mosovich and Tallaferro, "Studies on EEG," 219.

148. Mosovich and Tallaferro, "Studies on EEG," 219.

149. Letter from Dr. Alfred Kinsey to Dr. Abraham Mosovich, Feb. 29, 1952, Kinsey Correspondence Collection, box 13, Abraham Mosovich folder, KIL; Mosovich and Tallaferro, "Studies on EEG," 218.

150. Dr. Kinsey considered "the picture of epilepsy" to be "essential" to understandings of sexual climax. Letters from Dr. Alfred Kinsey to Dr. Abraham Mosovich, Feb. 29, 1952, and Dec. 1952, Kinsey Correspondence Collection, box 13, Abraham Mosovich folder, KIL.

151. Letter from Dr. Alfred Kinsey to Dr. Abraham Mosovich, Dec. 1952, Kinsey Correspondence Collection, box 13, Abraham Mosovich folder, KIL.

152. Kinsey, *Sexual Behavior in the Human Female*, 632.

153. Kinsey, *Sexual Behavior in the Human Female*, 632, 707.

154. Mosovich and Tallaferro, "Studies on EEG," 219; Lennox, *Epilepsy and Related Disorders*, 914.

155. Letter from Dr. Alfred Kinsey to Dr. Abraham Mosovich, Mar. 15, 1953, Kinsey Correspondence Collection, box 13, Abraham Mosovich folder, KIL.

156. Popular examples included authors and artists (Fyodor Dostoyevsky and Vincent Van Gogh), military leaders (Julius Caesar and Napoleon Bonaparte), and religious visionaries (Mohammed and Saint Paul).

157. Ernest, *Genius and Epilepsy*, 12.

158. Ernest, *Genius and Epilepsy*, 12.

159. Kroker, *Sleep of Others*.

160. Winter, *Memory*, 75–102.

161. Ernest, *Genius and Epilepsy*, 13.

162. "Geniuses Aid Tests," *New York Times*.

163. For more recent perspectives on EEG as a "post-hoc diagnostic procedure" used primarily to corroborate other observations: Delores Quinonez, "Common Applications of Electrophysiology (EEG) in the Past and Today: The Technologist's View," *Journal of Electroencephalography and Clinical Neurophysiology* 106, no. 2 (1998): 108–12. John Paul Leach and Rebecca O'Dwyer likewise argue that EEG "clouds the picture much of the time": John Paul Leach and Rebecca O'Dwyer, *Epilepsy Simplified* (Malta: Gutenberg Press, 2011), 56.

164. Lennox, *Science and Seizures*, 92–93.

165. For a parallel example of physicians questioning whether or not angiography provided the most meaningful representation of heart disease: David Jones, "Visions of a Cure: Visualization, Clinical Trials, and Controversies in Cardiac Therapeutics, 1968–1998," *Isis* 91, no. 3 (2000): 504–41.

166. After 1946, the number of EEG tests administered at the White Special School dropped precipitously, as the number of days the machine sat unused, broken, or gave inaccurate readings equated to "better than a 17% loss"; *Annual Report of the White Special School, 1947–1948*, 102, WPR; *Annual Report of the White Special School, 1949–1950*, 82, 84, WPR.

167. *Annual Report of the White Special School, 1948–1949*, 82, WPR.

168. *Annual Report of the White Special School, 1952–1953*, 12, WPR.

169. *Annual Report of the White Special School, 1952–1953*, WPR.

170. *Annual Report of the White Special School, 1955*, 14, WPR.

171. "Delinquents' Brains," *New York Times*.

172. Metzner advised Lennox of her difficulties in finding someone to operate and interpret EEG results at the White Special School. The school's nurse eventually filled the position. Letters from Alice Metzner to Dr. Lennox, July 16, 1941, and Oct. 28, 1941, Harvard Epilepsy Commission Papers, CLM.

173. Lennox, "Epilepsy: The Hopeful Disorder."

174. For example: letter from Dr. E. Schwade to Mr. Charles Crownhart, Jan. 27, 1949, State Medical Society Records, WHSL.

175. Kroker explains that emotion came to be understood as a function of adaptable group behavior rather than a static individual sign present in one's physiology. Kroker, "Washouts," 303–4.

176. Gibbs and Gibbs, *Atlas*, vol. 2, dedication page.

Chapter 3. Safe Citizens at Detroit's School for Children with Epilepsy

1. *Annual Report of the White Special School, 1945–1946*, 2, Detroit Public School Records, Acc #681, Walter P. Reuther Library (WPR), Archives of Labor and Urban Affairs, Wayne State University, Detroit, MI. This collection houses all of the annual reports of the White Special School mentioned in this chapter. For an earlier version of this research, see Rachel Elder, "Safe Seizures, Schoolyard Stoics, and the Making of Contained Citizens at Detroit's School for Epileptic Children, 1935–1956," *Journal of the History of Childhood and Youth* 7, no. 3 (2014): 430–61

2. *Annual Report of the White Special School, 1945–1946*, 10.

3. Aspects of citizenship were discussed with increasing frequency at the school throughout the 1940s, though terms such as "epileptic citizen" did not appear until the early 1950s. For example: "Building for Citizenship," *Annual Report of the White Special School, 1950–1951*, 87–88. On claims to able-bodiedness as a means of obtaining social citizenship in this period: David Serlin, "The Other Arms Race," in *The Disability Studies Reader*, 2nd ed., ed. Lennard J. Davis (New York: Routledge, 2006); Susan Burch, *Signs of Resistance: American Deaf Cultural History, 1900 to World War II* (New York: New York

University Press, 2002), 91; Kim E. Nielsen, *A Disability History of the United States* (Boston: Beacon, 2012).

4. Mona Gleason, *Normalizing the Ideal: Psychology, Schooling, and the Family in Postwar Canada* (Toronto: University of Toronto Press, 1999), 119–39.

5. Margaret A. Winzer, *From Integration to Inclusion: A History of Special Education in the 20th Century* (Washington, DC: Gallaudet University Press, 2009); Robert L. Osgood, *The History of Special Education: A Struggle for Equality in American Public Schools* (Westport, CT: Praeger, 2008).

6. For postwar characterization of epilepsy as "normal" and a "part-time handicap": "Epilepsy Isn't a Curse!" *Better Homes and Gardens*, Mar. 1948, 180–85, 180. For a popular characterization of epilepsy as an "invisible disability": Steven M. Spencer, "No Wonder Epileptics Are Bitter," *Saturday Evening Post*, Mar. 28, 1953. On the public construction of disability more generally: Rosemarie Garland Thomson, *Extraordinary Bodies: Figuring Disability in American Culture and Literature* (New York: Columbia University Press, 1997); Susan Schweik, *The Ugly Laws: Disability in Public* (New York: New York University Press, 2009).

7. On epilepsy as public and "out of the closet": Herbert Yahraes, *Epilepsy—The Ghost Is out of the Closet*, Public Affairs Pamphlet no. 98 (New York: Public Affairs Committee, 1944).

8. On medically "transmuted" diseases and experiences thereof: Chris Feudtner, *Bittersweet: Diabetes, Insulin, and the Transformation of Illness* (Chapel Hill: University of North Carolina Press, 2003).

9. On various forms of Cold War "containments": Elaine Tyler May, *Homeward Bound: American Families and the Cold War Era* (New York: Basic Books, 1988); Arnold R. Hirsch, "Containment on the Homefront: Race and Federal Housing Policy from the New Deal to the Cold War," *Journal of Urban History* 26 (2000): 158–89; David K. Johnson, *The Lavender Scare: The Cold War Persecution of Gays and Lesbians in the Federal Government* (Chicago: University of Chicago Press, 2009).

10. On the spaces and social engineering of childhood: Marta Gutman and Ning de Coninck-Smith, eds., *Designing Modern Childhoods: History, Space, and the Material Culture of Children* (New Brunswick: Rutgers University Press, 2008). On Cold War–era child safety and protection: Tarah Brookfield, *Cold War Comforts: Canadian Women, Child Safety, and Global Insecurity* (Waterloo: Wilfred Laurier University Press, 2012); Sharon Stephens, "Nationalism, Nuclear Policy and Children in Cold War America," *Childhood* 4 (1997): 103–23; Steven Mintz, *Huck's Raft: A History of American Childhood* (Cambridge, MA: Belknap, 2004), 275–309.

11. On nuclear and "duck and cover" culture in schools and classrooms: Paul Boyer, *By the Bomb's Early Light* (Chapel Hill: University of North Carolina Press, 1985). On postwar special education: Kathleen W. Jones, "Education for Children with Mental Retardation: Parent Activism, Public Policy, and Family Ideology in the 1950s," in *Mental Retardation in America: A Historical Reader,* ed. Stephen Noll and James W. Trent (New York: New York University Press, 2004); Winzer, *From Integration to Inclusion.*

12. On disability and criteria for social belonging: Kim Nielson, "Helen Keller and the

Politics of Civic Fitness," in *The New Disability History: American Perspectives*, ed. Paul K. Longmore and Lauri Umansky (New York: New York University Press, 2001); Allison C. Carey, *On the Margins of Citizenship: Intellectual Disability and Civil Rights in Twentieth-Century America* (Philadelphia: Temple University Press, 2009), 213; Joseph P. Shapiro, *No Pity: People with Disabilities Forging a New Civil Rights Movement* (New York: Random House, 1994).

13. I refer here to various forms of segregation in northern cities like Detroit that could be informal in nature but that were nevertheless present, both before and after the 1954 *Brown v. Board of Education* decision.

14. Request by Burt Shurly to Rackham Fund to support Handicapped Children Education, 1933, Detroit Public School Records, WPR.

15. On institutionalization and stigma: Ellen Dwyer, "Stigma and Epilepsy," *Transactions and Studies of the College of Physicians of Philadelphia*, 13, no. 4 (1991): 387–410; Ellen Dwyer, "Stories of Epilepsy, 1880–1930," in *Framing Disease: Studies in Cultural History*, ed. Charles Rosenberg and Janet Golden (New Brunswick: Rutgers University Press, 1992). On epilepsy and immigration: Douglas C. Baynton, "Defectives in the Land: Disability and American Immigration Policy, 1882–1924," *Journal of American Ethnic History* 24, no. 3 (2005): 31–44, 33.

16. Hildred Ileene Jarvis, "A Study of the Detroit Experimental School for Epileptic Children" (MA diss., Institute of Public and Social Administration, University of Michigan, 1937), 50; Edward M. Bridge, *Epilepsy and Convulsive Disorders in Children* (New York: McGraw-Hill, 1949).

17. Thomas D. Wood and Hugh Grant Rowell, *Health Supervision and Medical Inspection of Schools* (Philadelphia: W.B. Saunders, 1928), 383.

18. For epilepsy homeschooling statistics: Jarvis, "Detroit Experimental School," 9.

19. Jarvis, "Detroit Experimental School," 6.

20. Thomas Sugrue, *The Origins of the Urban Crisis: Race and Inequality in Postwar Detroit* (Princeton: Princeton University Press, 1996).

21. John W. Tenny, "A Study in Detroit's Special School Program," *Exceptional Children* 21, no. 5 (1955): 162–63.

22. Jarvis, "Detroit Experimental School," 18.

23. Jarvis, "Detroit Experimental School," 15–16.

24. *Annual Report of the White Special School, 1941*, 4; *Annual Report of the White Special School, 1944*, 8. On psychometrics and IQ testing in special education classes after 1916: Leila Zenderland, *Measuring Minds: Henry Herbert Goddard and the Origins of American Intelligence* (Cambridge: Cambridge University Press, 1998). Ellen Dwyer also documents the use of IQ testing in classes at the Craig Colony, noting that those students with results falling between 65 and 85 were considered the "most promising": Ellen Dwyer, "The State of the Multiply Disadvantaged: The Case of Epilepsy," in *Mental Retardation in America: A Historical Reader*, ed. Stephen Noll and James W. Trent (New York: New York University Press, 2004), 270.

25. Jarvis, "Detroit Experimental School," 52; *Annual Report of the White Special School, 1947–1948*, 78–79. Northern hospital schools and those for children with disabilities were often integrated. On segregated schools for children with disabilities: Carol

Linsenmeier and Jeff Moyer, "Reaching across the Divide: Visual Disability, Childhood, and American History," in *Children with Disabilities in America: A Historical Handbook and Guide,* ed. Philip Safford and Elizabeth Safford (Westport, CT: Greenwood, 2006); Susan Burch and Hannah Joyner, *Unspeakable: The Story of Junius Wilson* (Chapel Hill: University of North Carolina Press, 2007).

26. *Annual Report of the White Special School, 1947–1948,* 77; *Annual Report of the White Special School, 1955–1956,* 1.

27. Jarvis, "Detroit Experimental School," 4, 30.

28. O. P. Kimball and Don W. Gudakunst, "A Study of Epilepsy in Detroit," *Journal of the Michigan State Medical Society* (1936): 641–44; Jarvis, "Detroit Experimental School," 32. On concurrent clinical trials in children: Cynthia Connolly et al., "'A Startling New Chemotherapeutic Agent': Pediatric Infectious Disease and the Introduction of Sulfonamides at Baltimore's Sydenham Hospital," *Bulletin of the History of Medicine* 86, no. 1 (2012): 66–93.

29. Jarvis, "Detroit Experimental School," 37.

30. In particular, see *Laboratory Notes from Research Department, Park Davis & Co.,* 164 (July–Aug. 1937): 712, 735–36, Parke, Davis Research Laboratory Records, National Museum of American History Archives Center (NMAH), Smithsonian Institution, Washington, DC; *Annual Report of the White Special School, 1955,* 2. On the history of Dilantin: Lewis P. Rowland, *The Legacy of Tracy J. Putnam and H. Houston Merritt: Modern Neurology in the United States* (New York: Oxford University Press, 2008).

31. For example, Tridione was used at the White School between 1945 and 1947 before it was deemed unsafe and discontinued. *Annual Report of the White Special School, 1946–1947,* 102.

32. There was a waiting list for the school, and administrators noted that many parents were "desperate" to try new drugs that might stop or reduce their child's seizures. *Annual Report of the White Special School, 1943,* 2, 8.

33. *Annual Report of the White Special School, 1947–1948,* 82, 103; *Annual Report of the White Special School, 1941,* 4; *Annual Report of the White Special School, 1944,* 8.

34. Many White School teachers held master's degrees in special education from Wayne State University. They were almost exclusively female: *Annual Report of the White Special School, 1943,* 2. On the feminization of special education labor beginning in the Progressive era: Winzer, *From Integration to Inclusion,* 185–89.

35. Jarvis, "Detroit Experimental School," 11–14, 33; *Annual Report of the White Special School, 1955,* 3.

36. *Annual Report of the White Special School, 1955,* 3.

37. On nurses and data collection: Margarete Sandelowski, *Devices and Desires: Gender, Technology, and American Nursing* (Chapel Hill: University of North Carolina Press, 2000). On the changing nature of patient records and proliferation of patient data beginning in the interwar period: Joel Howell, *Technology in the Hospital: Transforming Patient Care in the Early Twentieth Century* (Baltimore: Johns Hopkins University Press, 1995).

38. *Annual Report of the White Special School, 1947–1948,* 96.

39. *Annual Report of the White Special School, 1955–1956,* 1.

40. *Annual Report of the White Special School, 1941,* 15.

41. For examples of postwar popular coverage of epilepsy, including of children with epilepsy: Robert Bassett, "The Problem of Epilepsy," *Hygeia* 24 (1946): 520–21; William Lennox, "Epilepsy: The Hopeful Disorder," *Parents' Magazine,* Jan. 1946, 124–27; "Some Plain English on Epilepsy," *Science Digest* 17 (1945): 82–83.

42. William G. Lennox, *Epilepsy Explained* (Boston: Laymen's League Against Epilepsy, 1940).

43. For instance, see Serlin, "The Other Arms Race."

44. William G. Lennox, "Rehabilitation of Epileptic Service Men," *American Journal of Psychiatry* 100 (Sept. 1943): 202–4, 204.

45. On "representational convergences" between wounded veterans and children: Seth Koven, "Remembering and Dismemberment: Crippled Children, Wounded Soldiers, and the Great War in Great Britain," *American Historical Review* 99, no. 4 (Oct. 1994): 1167–202; Lennox, "Epilepsy: The Hopeful Disorder," 124.

46. *Annual Report of the White Special School, 1945–1946,* 1. On medical optimism and media coverage in this era more generally: Bert Hansen, *Picturing Medical Progress from Pasteur to Polio: A History of Mass Media Images and Popular Attitudes in America* (New Brunswick: Rutgers University Press, 2009).

47. *Annual Report of the White Special School, 1945–1946,* 18, 122.

48. After its appearance in *Life* magazine, the White School received letters from thirty-four states and six foreign countries. Visitors included student nurses from Henry Ford Hospital, medical and social work students from Wayne State University, physicians, psychiatrists, drug company executives, and state directors of special education programs.

49. Critiques of "overprotection" among parents of children with epilepsy also began to show up in popular sources by the 1950s: Herb Bailey, "The Truth About Epilepsy," *Science Digest,* Sept. 1950, 73.

50. *Annual Report of the White Special School, 1947–1948,* 75. On the early twentieth-century rise of expert child-rearing advice and its acceleration after World War II: Anne Hulbert, *Raising America: Experts, Parents, and a Century of Advice about Children* (New York: Alfred A. Knopf, 2004). On the influence of psychology in regular and special education: Gleason, *Normalizing the Ideal;* Winzer, *From Integration to Inclusion,* 85–87, 140.

51. *Annual Report of the White Special School, 1943,* 3.

52. *Annual Report of the White Special School, 1949–50,* 88. On mother blame and their injurious devotion as a twentieth-century concern: *'Bad' Mothers: The Politics of Blame in Twentieth-Century America,* ed. Molly Ladd-Tayor and Lauri Umansky (New York: New York University Press, 1998); Rebecca Jo Plant, *Mom: The Transformation of Motherhood in Modern America* (Chicago: University of Chicago Press, 2010). On the "over-mothering" of children with other disabilities, including intellectual disabilities: Carey, *On the Margins of Citizenship.*

53. *Annual Report of the White Special School, 1941,* 4; *Annual Report of the White Special School, 1955–1956,* 2.

54. *Annual Report of the White Special School, 1947–1948,* 88; D. A. Pond, "Psychiatric Aspects of Epilepsy in Children," *British Journal of Psychiatry* 98 (1952): 404–10.

55. *Annual Report of the White Special School, 1945–1946,* 1.

56. A growing body of literature also discussed epilepsy as a "psychological handicap,"

referring to the anxiety associated with seizures and social disapproval: Lennox, "Epilepsy: The Hopeful Disorder," 126.

57. *Annual Report of the White Special School, 1947–1948*, 76; *Annual Report of the White Special School, 1949–1950*, 75; *Annual Report of the White Special School, 1953–1954*, 94. For concurrent examples of emerging parents' groups for children with disabilities after World War II: Jones, "Education for Children with Mental Retardation"; Carey, *On the Margins of Citizenship*, 105–33; Nielsen, *A Disability History of the United States*, 141–45.

58. *Annual Report of the White Special School, 1947–1948*, 76; *Annual Report of the White Special School, 1953–1954*, 94.

59. *Annual Report of the White Special School, 1949–1950*, 75.

60. *Annual Report of the White Special School, 1946–1947*, 98. For a similar but later case of individual, group, and family therapy for children with epilepsy: Zira de Fries and Sue Browder, "Group Therapy with Epileptic Children and Their Mothers," *Bulletin of the New York Academy of Medicine* 28, no. 4 (1952): 235–40.

61. *Annual Report of the White Special School, 1946–1947*, 99.

62. *Annual Report of the White Special School, 1947–1948*, 76.

63. *Annual Report of the White Special School, 1947–1948*.

64. *Annual Report of the White Special School, 1955*, 4.

65. Lennox, "Rehabilitation of Epileptic Service Men," 204. Dr. William Lennox appears to have become acquainted with the school in 1940 or 1941 through its quest for an EEG machine: letters from Alice Metzner to William G. Lennox, July–Dec. 1941, William Gordon Lennox Papers, B MS c 113, box 3, folder 9, Countway Library of Medicine (CLM), Center for the History of Medicine, Harvard University, Boston.

66. Lennox, "Rehabilitation of Epileptic Service Men," 204.

67. Mona Gleason discusses similar training in a larger number of postwar schools in efforts to produce "normal" citizens: Gleason, *Normalizing the Ideal*, 119–39.

68. On permissive play and child expressivity: Henry Jenkins, "Dennis the Menace, 'the All-American Handful,'" in *The Revolution Wasn't Televised: Sixties Television and Social Conflict*, ed. Lynn Spigel and Michael Curtin (New York: Routledge, 1997); Amy F. Ogata, "Creative Playthings: Educational Toys and Postwar American Culture," *Winterthur Portfolio* 39, no. 2/3 (June 1, 2004): 129–56; Roy Kovlovsky, "Adventure Playgrounds and Postwar Reconstruction," in *Designing Modern Childhoods*.

69. *Annual Report of the White Special School, 1948–1949*, 85.

70. *Annual Report of the White Special School, 1952–1953*, 2.

71. *Annual Report of the White Special School, 1955*, 1.

72. *Annual Report of the White Special School, 1951–1952*, 76.

73. On the employment of seventy-five "epileptics" at the River Rouge plant of the Ford Motor Company: Spencer, "No Wonder Epileptics Are Bitter," 26; Serlin, "The Other Arms Race," 49–65.

74. For examples of parents finding jobs for students, see the story of "Mr. H": *Annual Report of the White Special School, 1950–1951*, 97.

75. *Annual Report of the White Special School, 1943*, 4.

76. *Annual Report of the White Special School, 1955*, 3.

77. *Annual Report of the White Special School, 1950–1951*, 85.

78. On the expansion of child protectionism at home and in public: Brookfield, *Cold War Comforts*; Stephens, "Nationalism, Nuclear Policy and Children in Cold War America," 103–23.

79. *Annual Report of the White Special School, 1947–1948*, 82–84.

80. *Annual Report of the White Special School, 1947–1948*; Harry Sands, "Epilepsy and Accident Rates," *Epilepsia* 3 (1954): 122.

81. *Annual Report of the White Special School, 1947–1948*, 82.

82. *Annual Report of the White Special School, 1947–1948*.

83. *Annual Report of the White Special School, 1947–1948*, 83.

84. *Annual Report of the White Special School, 1947–1948*, 82. On schools and employers not wanting to assume responsibility for the risks and damages associated with seizures: Yahraes, *Epilepsy—The Ghost Is out of the Closet*, 22.

85. *Annual Report of the White Special School, 1947–1948*, 84.

86. Yahraes, *Epilepsy—The Ghost Is out of the Closet*, 22.

87. *Annual Report of the White Special School, 1946–1947*, 131.

88. "If Parents Only Knew the Tragedy of Preventable Falls," *Montreal Star*, May 23, 1950.

89. *Annual Report of the White Special School, 1946–1947*, 117.

90. Several scholars have noted that physical education was as much about "building character" as it was disciplining bodies: Susan A. Miller, *Growing Girls: The National Origins of Girls' Organizations in America* (New Brunswick: Rutgers University Press, 2007), 202. For a similar discussion regarding disability and sport, and regarding deaf communities in particular: Burch, *Signs of Resistance*, 75–83.

91. *Annual Report of the White Special School, 1949–1950*, 82.

92. *Annual Report of the White Special School, 1953–1954*, 91.

93. *Annual Report of the White Special School, 1946–1947*, 117–18.

94. For a parallel example, see also Burch, *Signs of Resistance*, 82.

95. *Annual Report of the White Special School, 1946–1947*, 118.

96. *Annual Report of the White Special School, 1949–1950*, 82.

97. *Annual Report of the White Special School, 1947–1948*, 98.

98. *Annual Report of the White Special School, 1949–1950*, 82–83.

99. *Annual Report of the White Special School, 1949–1950*.

100. *Annual Report of the White Special School, 1949–1950*.

101. *Annual Report of the White Special School, 1949–1950*.

102. *Annual Report of the White Special School, 1949–1950*; *Annual Report of the White Special School, 1946–1947*, 118. The White Special School competed with Jacoby and A. L. Holmes, which were two other schools for children with disabilities: *Annual Report of the White Special School, 1950–1951*, 81.

103. *Annual Report of the White Special School, 1945–1946*, 6.

104. *Annual Report of the White Special School, 1945–1946*, 14.

105. *Annual Report of the White Special School, 1945–1946*.

106. *Annual Report of the White Special School, 1949–1950*, 90; *Annual Report of the White Special School, 1945–1946*, 15; *Annual Report of the White Special School, 1947–1948*, 10.

107. Seizures were often blamed on uncooperative children and parents. Administrators noted that such children usually came from homes where "authority is divided" or where "money for drugs is not readily available," or from families who allegedly refused to wait at the Receiving Hospital for prescriptions: *Annual Report of the White Special School, 1953–1954*, 104.

108. In one case of a seizure on a field trip, staff later agreed that the student should not have gone on the trip anyway since he had experienced a "hard seizure" both the night and morning beforehand: *Annual Report of the White Special School, 1949–1950*, 90.

109. *Annual Report of the White Special School, 1950–1951*, 1.

110. *Annual Report of the White Special School, 1945–1946*, 10; *Annual Report of the White Special School, 1946–1947*,103; *Annual Report of the White Special School, 1947–1948*, 98; *Annual Report of the White Special School, 1948–1949*, 88.

111. *Annual Report of the White Special School, 1947–1948*, 77.

112. *Annual Report of the White Special School, 1947–1948*, 94.

113. *Annual Report of the White Special School, 1951–1952*, 76.

114. *Annual Report of the White Special School, 1943*, 1.

115. *Annual Report of the White Special School, 1955*, 5.

116. On the mainstreaming of special education: Osgood, *The History of Special Education*, 114–118.

117. On discourses of "normalcy" in various aspects of postwar culture: Anna Creadick, *Perfectly Average: The Pursuit of Normality in Postwar America* (Amherst: University of Massachusetts Press, 2010); Sarah Igo, *The Averaged American: Surveys, Citizens, and the Making of a Mass Public* (Cambridge: Harvard University Press, 2006); David Serlin, *Replaceable You: Engineering the Body in Postwar America* (Chicago: University of Chicago Press, 2002); Mary Louise Adams, *The Trouble with Normal: Postwar Youth and the Making of Heterosexuality* (Toronto: University of Toronto Press, 1997).

118. *Annual Report of the White Special School, 1948–1949*, 74–75.

119. Lennox, "Rehabilitation of Epileptic Service Men," 203; Spencer, "No Wonder Epileptics Are Bitter," 53.

120. *Annual Report of the White Special School, 1950–1951*, 85.

121. *Annual Report of the White Special School, 1952–1953*, 4; *Annual Report of the White Special School, 1955*, 1; *Annual Report of the White Special School, 1947–1948*, 94.

122. *Annual Report of the White Special School, 1950–1951*, 87.

123. *Annual Report of the White Special School, 1951–1952*, 79.

124. Discussions of marriage, children, and employment are ubiquitous in articles on epilepsy in this era: Lennox, "Epilepsy: The Hopeful Disorder," 127; Matthias M. Krenn, "Social Outcast—Age 9," *Today's Health* 29, no. 8 (1951): 28–30, 30.

125. *Annual Report of the White Special School, 1951–1952*, 79.

126. *Annual Report of the White Special School, 1949–1950*, 82–83.

127. *Annual Report of the White Special School, 1953–1954*, 90.

128. *Annual Report of the White Special School, 1953–1954*, 104.

129. *Annual Report of the White Special School, 1944–1945*; *Annual Report of the White Special School, 1948–1949*.

130. *Annual Report of the White Special School, 1948–1949.*

131. On disability and the "perpetual performance of overcoming": Nielson, "Helen Keller and the Politics of Civic Fitness," 269.

132. *Annual Report of the White Special School, 1947–1948,* 94.

133. "Epilepsy," *Life,* June 3, 1946, 129–34. Note that such publicity only slightly predated forthcoming publications that would expose the abysmal conditions of various psychiatric hospitals across the country. For instance, see Allison Carey's discussion of Frank Leon Wright's 1947 book, *Out of Sight, Out of Mind,* which in detailing such conditions served as a consciousness-raising precursor to deinstitutionalization and the disability rights movement: Carey, *On the Margins of Citizenship,* 92–93.

134. In September 1956, the school became the White School for Crippled Children. *Annual Report of the White Special School, 1955–1956,* 8.

Chapter 4. Mobility and Motor Control

1. Report by Orren P. Olsen, Eastern Conference of Motor Vehicle Administrators, n.d. (c. 1948), folder 6, Box 88, MS 843, State Medical Society of Wisconsin (SMSW) Records, Wisconsin Historical Society Library (WHSL), University of Wisconsin, Madison, WI.

2. Olsen, Eastern Conference of Motor Vehicle Administrators, n.d.

3. W. E. Carter, "The Epileptic Driver," *Journal of Mental and Nervous Disease* 99, no. 5 (1944): 573–75.

4. H. D. Fabing and R. L. Barrow, *Epilepsy and the Law: A Proposal for Legal Reform in the Light of Medical Progress* (New York: Hoeber-Harper, 1956), 36.

5. Jeremy Packer, *Mobility without Mayhem: Safety, Cars, and Citizenship* (Durham: Duke University Press, 2008); Cotton Seiler, *Republic of Drivers: A Cultural History of Automobility in America* (Chicago: University of Chicago Press, 2008).

6. On Dilantin: Lewis P. Rowland, *The Legacy of Tracy Putnam and H. Houston Merritt: Modern Neurology in the United States* (New York: Oxford University Press, 2008). On "disease transmutation" and chronic illness: Chris Feudtner, *Bittersweet: Diabetes, Insulin, and the Transformation of Illness* (Chapel Hill: University of North Carolina Press, 2007).

7. L. E. Himler, "Epilepsy as a Traffic Hazard," *Journal of the Michigan State Medical Society* 40 (1941): 707–10, 710.

8. For instance: Judith Walzer Leavitt, *Typhoid Mary: Captive to the Public's Health* (Boston: Beacon, 1996); Alan Kraut, *Silent Travelers: Germs, Genes, and the Immigrant Menace* (Baltimore: Johns Hopkins University Press, 1994); Charles Rosenberg, *The Cholera Years: The United States in 1832, 1849, and 1866* (Chicago: University of Chicago Press, 1987); Kara Swanson, *Banking the Body: Blood, Milk, and Sperm in Modern America* (Cambridge: Harvard University Press, 2014).

9. Leavitt, *Typhoid Mary,* 1–4.

10. Barron H. Lerner, *One for the Road: Drunk Driving Since 1900* (Baltimore: Johns Hopkins University Press, 2011); Amy Gangloff, "Medicalizing the Automobile: Public Health, Safety, and American Culture, 1920–1967" (PhD diss., State University of New York at Stony Brook, 2006); Lee Vinsel, "'Safe Driving Depends on the Man at the Wheel': Psychologists and the Subject of Auto Safety, 1920–55," *Osiris* 33, no. 1 (2018): 191–209.

11. Sam Bagby, "Diplomas for Bad Drivers," *Los Angeles Times,* Sept. 44, 1938, G11;

"They Ask a Drivers' License Law," *Minneapolis Star*, Oct. 15, 1931, 16; Carroll Mealey, "Motor Bureau Has Broad Plans," *New York Times*, Oct. 13, 1940, 150.

12. Concerns about driver fitness throughout the interwar period focused mostly on the intelligence and age of the driver. Alan Canty, "The Case Study Method of Rehabilitating Drivers," *Journal of Social Psychology* 12, no. 2 (1940): 271–78.

13. Unlike old age or poor vision, drivers experiencing diabetic episodes and seizures were said to act "in a Dr. Jekyll and Mr. Hyde capacity" on the road. Bagby, "Diplomas for Bad Drivers."

14. Peter Norton, *Fighting Traffic: The Dawn of the Motor Age in the American City* (Cambridge: MIT Press, 2008).

15. Susan Burch, *Signs of Resistance: American Deaf Cultural History, 1900 to World War II* (New York: New York University Press, 2002), 155–65.

16. Packer, *Mobility without Mayhem*, 33–68.

17. W. E. Carter and R. W. Harvey, "Epilepsy—A Reportable Disease," *California State Department of Public Health Weekly Bulletin* 18 (Aug. 1939): 109.

18. Carter, "Epileptic Driver," 573.

19. Carter and Harvey, "A Reportable Disease," 109.

20. Newspapers of the mid-1930s, for instance, noted that "police have reported many cases of automobile accidents due to sudden attacks by epileptic drivers." Morris Fishbein, "Epileptic Must Avoid Hazardous Jobs," *Chicago Press*, Dec. 22, 1934, A10.

21. On interwar crash consciousness: Amy Gangloff, "Safety in Accidents: Hugh DeHaven and the Development of Crash Injury Studies," *Technology and Culture* 54, no. 1 (2013): 40–61; David Blanke, *Hell on Wheels: The Promise and Peril of America's Car Culture, 1900–1940* (Lawrence: Kansas University Press, 2007).

22. On public backlash to car dominance and accidents after 1930: Norton, *Fighting Traffic*.

23. Norton, *Fighting Traffic*; Vinsel, "Safe Driving Depends."

24. "The Human Factor in Motor Vehicle Accidents: Physical or Mental Disabilities," *JAMA* 116 (1941): 2402–3. On educating drivers: Albert Whitney, *Man and the Motor Car* (New York: J. J. Little and Ives, 1936).

25. Earl J. Reeder, "Locating Automatic Traffic Signals," Report of the Public Safety Division, Department of Publicity (Chicago: National Safety Council, Apr. 25, 1926); "Committee on Night Traffic Hazards: Visibility vs. Traffic Accidents," 1039 Report to Street and Highway Traffic Section (Chicago: National Safety Council, Oct. 18, 1939); "Speed Driving Target of Bills. State Limits Urged," *Milwaukee Journal*, Mar. 4, 1949.

26. Fabing and Barrow, *Epilepsy and the Law*, 57.

27. Gangloff, "Safety in Accidents"; Lerner, *One for the Road*.

28. William Lennox, *Epilepsy and Related Disorders* (Boston: Little, Brown, 1960), 987.

29. "The Human Factor," *JAMA*, 2402; Carter, "Epileptic Driver," 573; Vinsel, "Safe Driving Depends."

30. Ellen Dwyer, "Stories of Epilepsy, 1880–1930," in *Framing Disease: Studies in Cultural History,* ed. Charles Rosenberg and Janet Golden (New Brunswick: Rutgers University Press, 1992); Douglas Baynton, "Defectives in the Land: Disability and American Immigration Policy, 1882–1924," *Journal of American Ethnic History* 24, no. 3 (2005): 31–44.

31. "The Human Factor," *JAMA,* 2402.

32. W. E. Carter and R. W. Harvey, "Epilepsy: A Hazardous Disease," *California and Western Medicine* 56, no. 5 (1942): 294–95, 294.

33. Himler, "Epilepsy as Traffic Hazard," 710.

34. Lennox, *Epilepsy and Related Disorders,* 987.

35. Carter and Harvey, "A Hazardous Disease," 295.

36. Himler, "Epilepsy as Traffic Hazard," 707–8.

37. Letter from Dr. Holloway to Dr. Crownhart, Sept. 4, 1947, MSS 843, box 88, folder 6, SMSW Records, WHSL; G. H. Monrad-Krohn, "Epilepsy and Motoring," *Epilepsia* II 1, no. 3 (1939): 192–95.

38. Carter, "Epileptic Driver," 573.

39. Carter and Harvey, "A Reportable Disease," 109.

40. Himler, "Epilepsy as Traffic Hazard," 707.

41. Carter and Harvey, "A Hazardous Disease," 294.

42. On tuberculosis and syphilis, respectively: Cynthia Connolly, *Saving Sickly Children: The Tuberculosis Preventorium in American Life, 1909–1970* (New Brunswick: Rutgers University Press, 2008); Allan Brandt, *No Magic Bullet: A Social History of Venereal Disease in the United States Since 1880* (New York: Oxford University Press, 1985).

43. Carter, "Epileptic Driver," 575.

44. William Lennox, "Rehabilitation of Epileptic Service Men," *American Journal of Psychiatry* 100 (1943): 202–4.

45. Himler, "Epilepsy as Traffic Hazard," 707–8.

46. Himler, "Epilepsy as Traffic Hazard," 707–8.

47. Carter and Harvey, "A Reportable Disease," 109.

48. Himler, "Epilepsy as Traffic Hazard," 707–8.

49. "The Human Factor," *JAMA,* 2403. A parallel example of medical screenings for invisible "pathologies" is that of the "sexual invert" in the 1930s: Jennifer Terry, *An American Obsession: Science, Medicine, and Homosexuality in Modern Society* (Chicago: University of Chicago Press, 1999).

50. Carter and Harvey, "A Reportable Disease," 109.

51. Carter, "Epileptic Driver," 573.

52. Himler, "Epilepsy as Traffic Hazard," 709.

53. Carter, "Epileptic Driver," 574.

54. Himler, "Epilepsy as Traffic Hazard," 708.

55. Lowell S. Selling, "Convulsive Disorders and the Automobile Driver," *American Journal of Psychiatry* 99, no. 6 (1943): 869–71, 869.

56. H. A. Heise et al. "Report of Committee to Study Problems of Motor Vehicle Accidents," *JAMA* 119, no. 8 (1942): 653–55.

57. Heise et al., "Problems of Motor Vehicle Accidents," 653.

58. Heise et al., "Problems of Motor Vehicle Accidents," 653.

59. Heise et al., "Problems of Motor Vehicle Accidents," 654.

60. On disabled workers during World War II more generally: David Serlin, "The Other Arms Race," in *The Disability Studies Reader,* 2nd ed., ed. Lennard J. Davis (New York: Routledge, 2006).

61. Carter, "Epileptic Driver," 575.

62. Margot Canaday, *The Straight State: Sexuality and Citizenship in Twentieth-Century America* (Princeton: Princeton University Press, 2009), 91–136; Carter, "Epileptic Driver," 575.

63. Heise et al., "Problems of Motor Vehicle Accidents," 653–54.

64. "Informants" were typically friends, relatives, and physicians of people with epilepsy. Selling, "Convulsive Disorders," 870.

65. Carter and Harvey, "A Hazardous Disease," 295.

66. Himler, "Epilepsy as Traffic Hazard," 710.

67. Himler, "Epilepsy as Traffic Hazard," 710.

68. Mark Albert Glaser and Martin Irons, "Is the Epileptic a Safe Motor Vehicle Driver?" *Journal of Mental and Nervous Diseases* 103, no. 1 (1946): 21–36, 25.

69. Himler, "Epilepsy as Traffic Hazard," 709; Carter, "Epileptic Driver," 574.

70. Letter from Dr. Holloway to Dr. Crownhart, Sept. 4, 1947, SMSW Records.

71. Eugene Canudo, "Combatting Auto Accidents: State Inspection of Vehicles Favored with Re-Examination of Drivers," *New York Times*, Nov. 25, 1948, 30.

72. Packer, *Mobility without Mayhem*, 28.

73. As reported in letter from John Thomas of the Motor Vehicle Department to Mr. Charles Crownhart, Aug. 17, 1956, SMSW Records.

74. Letter from Dr. Crownhart to Dr. Holloway, Sept. 11, 1947, SMSW Records, WHSL.

75. Letter from Dr. E. Schwade to Mr. Charles Crownhart, Mar. 2, 1949, SMSW Records, WHSL.

76. Himler, "Epilepsy as Traffic Hazard," 707–8.

77. Letter from Dr. E. Schwade to Mr. Charles Crownhart, June 7, 1949, SMSW Records, WHSL.

78. Olsen, Eastern Conference of Motor Vehicle Administrators, n.d.

79. Glaser and Irons, "Is the Epileptic Safe?" 24.

80. Glaser and Irons, "Is the Epileptic Safe?" 24.

81. Glaser and Irons, "Is the Epileptic Safe?" 31.

82. Olsen, Eastern Conference of Motor Vehicle Administrators, n.d. On contemporaneous safety councils: Norton, *Fighting Traffic*.

83. Letter from Dr. E. Schwade to Mr. Charles Crownhart, Mar. 2, 1949, SMSW Records, WHSL.

84. Letter from Schwade to Crownhart, Mar. 2, 1949.

85. Selling, "Convulsive Disorders," 871.

86. William Lennox, "The Epileptic Driver," *JAMA* 137, no. 3 (May 1948): 322.

87. Letter from John Thomas of the Motor Vehicle Department to Mr. Charles Crownhart, Aug. 17, 1956, SMSW Records, WHSL; Olsen, Eastern Conference of Motor Vehicle Administrators. n.d.

88. Glaser and Irons, "Is the Epileptic Safe?" 32.

89. For early revisionist depictions of epilepsy in the 1940s: Herbert Yahraes, *Epilepsy—The Ghost Is out of the Closet*, Public Affairs Pamphlet no. 98 (New York: Public Affairs Committee, 1944).

90. Lennox, "Rehabilitation of Epileptic Service Men."

91. Letter from Dr. E. Schwade to Mr. Charles Crownhart, Mar. 2, 1949, SMSW Records, WHSL.

92. Letter from Schwade to Crownhart, Mar. 2, 1949.

93. Letter from Dr. E. Schwade to Mr. Charles Crownhart, Dec. 13, 1948, SMSW Records, WHSL.

94. Letter from Dr. E. Schwade to Mr. Charles Crownhart, Oct. 17, 1948, and Dec. 13, 1948, SMSW Records, WHSL.

95. Letter from Dr. E. Schwade to Mr. Charles Crownhart, Jan. 27, 1949, SMSW Records, WHSL.

96. Letter from John Thomas to Mr. Charles Crownhart, Aug. 17, 1956, SMSW Records, WHSL.

97. Letter from Thomas to Crownhart, Aug. 17, 1956.

98. Fabing and Barrow, *Epilepsy and the Law*, 127.

99. "Speed Driving," *Milwaukee Journal*, Mar. 4, 1949.

100. On the supreme importance of "reliability" in this history: "Statement of Francis M. Forster, MD," enclosed in letter from Wisconsin Epilepsy Association to Mr. Lowell Jackson, Apr. 2, 1979, Secretary Subject Files, 1964–1966, box 23, Wisconsin Department of Transportation (WDT), WHSL.

101. All cases have been anonymized by the author. Case B.R. "1954," Administrator's General Correspondence, 1949–1979, series 2328, Wisconsin Division of Motor Vehicles (WDMV), WHSL.

102. Case B.R. "1954."

103. Case B.R. "1954."

104. Case K.H. "1954."

105. Case I.J. "1954."

106. Case H.J. "May 1951."

107. Case T.J. "October 1951."

108. On patent medicines: Case B.R. "1949"; Case R.R. "May 1950." On electrical treatments: Case O.G. "May 1950." On changing doctors: Case T.W. "1949."

109. Case H.C. "May 1951"; Case T.W. "October 1951."

110. Case M.L. "1949."

111. Case T.P. "October 1951."

112. Respectively, Cases I.F. "1949," O.R. "May 1950," and N.J. "October 1950."

113. On women and "stress": Cases K.H. "October 1950" and V.S. "October 1951." On traumatic epilepsy: Cases M.P. and S.R. "October 1951" and T.W. "October 1950."

114. Cases P.L. and L.C. "October 1950."

115. On concussion: Case T.W. "October 1950." On battle fatigue: Cases I.J. and I.R. "May 1950."

116. A third of cases in November 1949 were no-shows. Reasons for missing hearings included inability to miss work, lack of transportation, and disinterest in a license. See "Nov. 1949," Administrator's General Correspondence, 1949–1979, series 2328, WDMV, WHSL.

117. Case S.A. "1954."

118. Case E.J. "May 1950."

119. Case S.F. "May 1951."

120. Case S.L. "May 1950."

121. Case I.N. "October 1951."

122. Letter from Dr. E. Schwade to Mr. Charles Crownhart, Mar. 2, 1949, SMSW Records, WHSL.

123. Case N.M. "1954."

124. Case I.R. "1954."

125. Cases L.J. "May 1950" and I.G. "1954."

126. Cases I.J. and I.R. "May 1950."

127. Case T.R. and W.W. "1954."

128. Case M.S. "1954."

129. Cases H.R. "May 1950" and D.E. "1949."

130. Lennox, *Epilepsy and Related Disorders,* 990. On "women drivers": Packer, *Mobility without Mayhem*, 33–54. On women's place in the suburban domestic order: Elaine Tyler May, *Homeward Bound: American Families and the Cold War Era* (New York: Basic Books, 1988).

131. Cases S.L. "1954" and S.M. "1949," WDMV, WHSL.

132. Case H.B. "1954."

133. Case E.G. "1954."

134. Case E.G. "1954."

135. Case F.E. "1954."

136. Case M.C. "1954."

137. Census data for 1940, 1950, and 1960 shows the percentage of white inhabitants in Wisconsin to be 99.2, 98.8, and 97.6 percent, respectively. Campbell Gibson and Kay Jung, "Table 64—"Wisconsin, Race and Hispanic Origin: 1820–1990," in "Population Division: Historical Census Statistics on Population Totals by Race 1790 to 1990, and by Hispanic Origin, 1970 to 1990, for the United States, Regions, Divisions, and States," Working Paper No. 56 (Washington, DC: US Census Bureau, Population Division, Sept. 2002), https://www2.census.gov/library/working-papers/2002/demo/pop-twps0056/table64.pdf.

138. There is no explicit discussion of class, race, or ethnicity in the records of Wisconsin's epilepsy hearings. The age, gender, marital status, and occasionally occupation of the applicant are included.

139. On race, gender, and family role as it relates to epilepsy and issues of work and driving, see the discussion of "Mr. and Mrs. Washington" in Horace B. Eldred, "Epilepsy in Adult Males, Impact on Wives" (MA thesis, Faculty of the School of Social Work, University of Southern California, 1959), 12, 17, 31, 37–39.

140. For a later example of the restrictions applicants could face, see Transcript of Medical Review Board, "Meeting with D.Mc. (applicant anonymized)," May 1978, Office of General Counsel, Attorney Opinions, 1967–86 (series 2354), Part 1 (1967–), box 15, folder 18 (Epilepsy Affidavits), WDT, WHSL.

141. The twelve states were Arizona, Delaware, Indiana, Kansas, Kentucky, Michigan, North Carolina, Ohio, Pennsylvania, Virginia, Washington, and Wisconsin. Fabing and Barrow, *Epilepsy and the Law,* 41.

142. Lennox, *Epilepsy and Related Disorders*, 988.

143. "Epileptics to Meet, Map New Campaign," *Milwaukee Sentinel*, Sept. 17, 1954; "Allow Marriage of Epileptics?" *Milwaukee Journal*, n.d., SMSW Records, WHSL; "Fight Waged for Epileptics: They Aren't Mentally Ill, Milwaukee Doctor Tells Legislators" *Milwaukee Journal*, May 5, 1955.

144. H. D. Fabing and R. L. Barrow, "Restricted Driver's Licenses to Controlled Epileptics: A Realistic Approach to a Problem of Highway Safety," *UCLA Law Review* 2 (1954–55): 500–14, 510.

145. "Plea for Epileptics: End of Driving Curbs on Afflicted to Be Sought," *New York Times*, Nov. 29, 1955.

146. Fabing and Barrow, *Epilepsy and the Law*, 35.

147. Fabing and Barrow, *Epilepsy and the Law*, 36; Fabing and Barrow, "Restricted Drivers' Licenses," 509.

148. "Epileptics to Meet," *Milwaukee Sentinel*; "Plea for Epileptics," *New York Times*.

149. Lennox, *Epilepsy and Related Disorders*, 988.

150. Lennox, *Epilepsy and Related Disorders*, 988.

151. Fabing and Barrow, "Restricted Drivers' Licenses," 501.

152. Selling, "Convulsive Disorders," 870.

153. Fabing and Barrow, *Epilepsy and the Law*, 35.

154. "Army Enlistment Open to Controlled Epileptic," *Milwaukee Journal*, Feb. 18, 1953; Robert Root, "Epileptics as Drivers," *New York Times*, June 29, 1958, E8.

155. For a letter of a parent whose child was struck by an "epileptic driver": Marshall Shields, "Epileptic Drivers," *Chicago Daily Tribune*, Nov. 30, 1957, 12; Fred Bodsworth, "License to Murder?" *Maclean's*, Oct. 15, 1950, 14–15, 60–62.

156. Lennox, *Epilepsy and Related Disorders*, 993.

157. Lennox, *Epilepsy and Related Disorders*, 994; "Court Orders Epileptic to Face Trial in Car Death," *Newsday*, Nov. 30, 1956, 4; Fabing and Barrow, *Epilepsy and the Law*, 91.

158. According to a Canadian source in 1966, "Almost always they have some kind of warning an attack is coming": "Epileptic Driver Has Seizure, Four Injured," *Globe and Mail*, Sept. 30, 1966, 5.

159. Glaser and Irons, "Is the Epileptic Safe?" 26.

160. Lennox, *Epilepsy and Related Disorders*, 993.

161. Lennox, *Epilepsy and Related Disorders*, 933.

162. For an earlier verdict of non-guilt by a young female driver: "Ban on Epileptic Driver Urged in Police Killing," *Chicago Daily Tribune*, Oct. 4, 1940, 33. An even earlier case, reported in a sympathetic manner, referred to a driver with epilepsy who crashed as "the unfortunate driver." "Epileptic Fit Strikes Driver: Unfortunate Man Not Badly Hurt as Car Crashes," *Los Angeles Times*, Aug. 7, 1928, 12.

163. Lennox, *Epilepsy and Related Disorders*, 991–93.

164. Lennox, "Epileptic Driver."

165. Letter from Dr. E. Schwade to Mr. Charles Crownhart, March 9, 1949, SMSW Records, WHSL.

166. Lennox, *Epilepsy and Related Disorders*, 990.

167. "The Human Factor," *JAMA*, 2402; Lennox, *Epilepsy and Related Disorders,* 994.

168. Lerner, *One for the Road*.

169. "The Human Factor," *JAMA,* 2402; Lennox, *Epilepsy and Related Disorders,* 994.

170. W. L. Neustatter, *The Mind of the Murderer* (New York: Philosophical Library, 1957), 70–83.

171. Lennox, *Epilepsy and Related Disorders*, 991–93. Others, well into the later twentieth century, continued to see seizures as a "failure" on the part of the patient, including failure to take medication properly or to get proper rest: "Dear Abby: Taking Anonymous Route to Curb Epileptic Driver," *Chicago Tribune*, Mar. 19, 1981.

172. Letter from John P. Desmond, J.D., to Mr. Charles Crownhart, Sept. 9, 1959, SMSW Records, WHSL.

173. Fabing and Barrow, "Restricted Drivers' Licenses," 506.

174. Lennox, *Epilepsy and Related Disorders*, 991.

175. Root, "Epileptics as Drivers," E8.

176. Letter from Dr. H. C. Marsh to the State Medical Society, Aug. 4, 1959; Dr. R. Ramlow to the State Medicine Society, Feb. 25, 1957, SMSW Records, WHSL; Homer Bigart, "Epileptics Face Test on Driving," *New York Times,* Nov. 12, 1955, 21.

177. Letter from John P. Desmond, J.D., to Mr. Charles Crownhart, Sept. 9, 1959, SMSW Records, WHSL.

178. Fabing and Barrow, "Restricted Drivers' Licenses," 506; letter from Dr. R. Ramlow to the SMS, Feb. 25, 1957, SMSW Records, WHSL.

179. Letter from Dr. E. Schwade to Mr. Charles Crownhart, March 2, 1949, SMSW Records, WHSL.

180. Note that in 1963, eleven states denied driver's permits to people with epilepsy altogether: Alabama, Delaware, Indiana, Kentucky, Michigan, Minnesota, North Carolina, Ohio, Pennsylvania, Virginia, and Washington. "Epileptic Drivers Barred," *Hartford Courant*, Feb. 1, 1963, 17.

181. Letter from Mrs. James Hanson to Mr. Lowell Jackson, July 7, 1979, WDT, WHSL.

182. Letter from Mrs. James Hanson to Mr. Lowell, July 7, 1979, WDT, WHSL.

183. Letter from Mrs. James Hanson to Mr. Lowell, July 7, 1979, WDT, WHSL.

184. Wisconsin Epilepsy Association to Mr. Lowell Jackson, April 2, 1979, WDT, WHSL.

185. Wisconsin Epilepsy Association to Mr. Lowell Jackson, April 2, 1979, WDT, WHSL.

186. Transcript of Medical Review Board, "Meeting with D.Mc.," May 1978, WDT, WHSL.

187. For example, see expanding military and workplace inclusion in this era: "Army Enlistment Open to Controlled Epileptic," *Milwaukee Journal*, Feb. 18, 1953.

Chapter 5. The Operable Brain

1. First published in Wilder Penfield, "The Cerebral Cortex in Man: 1. The Cerebral Cortex and Consciousness," *Archives of Neurology and Psychiatry* 40, no. 3 (1938): 431. Note that an earlier version of the research in this chapter appeared as Rachel Elder, "Speaking Secrets: Epilepsy, Neurosurgery, and Patient Testimony in the Age of the Explorable Brain," *Bulletin of the History of Medicine* 89, no. 4 (2015): 761–89.

2. All patients have been anonymized to protect their identities. In most cases, patients have been given pseudo-initials, though in some cases I have avoided initials

entirely. The year listed alongside patient records reflects the year the patient's file was created and is usually the same as the year the patient was first operated on. All patient and neuropathology records are housed at the Montreal Neurological Institute-Hospital, also called the Montreal Neurological Institute, or MNI.

3. Wilder Penfield, ed., *The Mysteries of the Mind: A Critical Study of Consciousness and the Human Brain* (Princeton: Princeton University Press, 1975).

4. Wilder Penfield, "Activation of the Record of Human Experience: Summary of the Lister Oration Delivered at the Royal College of Surgeons of England on 27th April 1961," *Annals of the Royal College of Surgeons of England* 29, no. 2 (1961): 77–84.

5. Alison Winter, *Memory: Fragments of a Modern History* (Chicago: University of Chicago Press, 2012); Cathy Gere, "'Nature's Experiment': Epilepsy, Localization of Brain Function and the Emergence of the Cerebral Subject," in *Neurocultures: Glimpses into an Expanding Universe*, ed. Francisco Ortegon and Fernando Vidal (Frankfurt: Peter Lang, 2006).

6. Popular histories generally focus on Penfield's discoveries and the institutional history of the MNI: Jefferson Lewis, *Something Hidden: A Biography of Wilder Penfield* (Toronto: Doubleday Canada, 1981); Wilder Penfield, *No Man Alone: A Neurosurgeon's Life* (Boston: Little, Brown, 1977). Histories of science and medicine mostly examine the experts, practices, and content of the science: William Feindel, "Epilepsy Surgery in Canada," in *Textbook of Epilepsy Surgery,* ed. Hans O. Luders (London: Informa, 2008); Delia Gavrus, "Men of Dreams and Men of Action: Neurologists, Neurosurgeons, and the Performance of Professional Identity, 1920–1950," *Bulletin of the History of Medicine* 85, no. 1 (2011): 57–92; Winter, *Memory: Fragments of a Modern History*; Gere, "'Nature's Experiment.'"

7. Though neither study systematically includes patients, both Alison Winter and Cathy Gere acknowledge the role of Penfield's patients in producing knowledge about the brain. Gere argues that the "cerebral subject" is an extrapolation of the epileptic brain, while Winter highlights the importance of patient feedback during surgical stimulations: Gere, "'Nature's Experiment'"; Winter, *Memory: Fragments of a Modern History*.

8. This is in direct contrast to the "risky experiments of uncertain value" performed on institutionalized epilepsy sufferers during this period: Ellen Dwyer, "Neurological Patients as Experimental Subjects," in *The Neurological Patient in History,* ed. Stephen Jacyna and Stephen Casper (Rochester, NY: University of Rochester Press, 2012).

9. On psychosurgery: Jack D. Pressman, *Last Resort: Psychosurgery and the Limits of Medicine* (Cambridge: Cambridge University Press, 1998).

10. On surgery as a modern "fix": Thomas Schlich, "The Technological Fix and the Modern Body: Surgery as a Paradigmatic Case," in *The Cultural History of the Human Body,* vol. 6, ed. Linda Kalof and William Bynum (London: Berg, 2010). On perceptions of epilepsy in the early twentieth century: Ellen Dwyer, "Stories of Epilepsy, 1880–1930," in *Framing Disease: Studies in Cultural History,* ed. Charles Rosenberg and Janet Golden (New Brunswick: Rutgers University Press, 1992).

11. Stephen Jacyna and Stephen Casper, *The Neurological Patient in History* (Rochester, NY: University of Rochester Press, 2012).

12. Fernando Vidal, "Brainhood, Anthropological Figure of Modernity," *History of the Human Sciences* 22, no. 1 (2009): 5–36; Jacyna and Casper, *The Neurological Patient*, 1, 9;

Laura Salisbury and Andrew Shail, *Neurology and Modernity: A Cultural History of Nervous Systems, 1800–1950* (London: Palgrave Macmillan, 2010).

13. On patient H.M.: Suzanne Corkin, *Permanent Present Tense: The Man with No Memory, and What He Taught the World* (New York: Allen Lane, 2013); Max Stadler, "The Neurological Patient in History," in Jacyna and Casper, *The Neurological Patient*.

14. For the parallel example of aphasia: Stephen Jacyna, *Lost Words: Narratives of Language and the Brain, 1825–1926* (Princeton: Princeton University Press, 2000).

15. Roy Porter, "The Patient's View: Doing Medical History from Below," *Theory and Society* 14 (1985): 175–98.

16. J.V. pathological report, *Laboratory of Neuropathology: Neurosurgical Reports*, 1936, Montreal Neurological Institute (MNI), Montreal, QC.

17. J.V. pathological report, *Laboratory of Neuropathology*, 1936.

18. J.V. pathological report, *Laboratory of Neuropathology*, 1936.

19. J.V. pathological report, *Laboratory of Neuropathology*, 1936.

20. J.V. pathological report, *Laboratory of Neuropathology*, 1936.

21. J.V. patient record, Montreal Neurological Institute (MNI), Montreal, QC.

22. Leo M. Davidoff, "A Visit to Professor Foerster's Clinic in Breslau with Special Observation of His Treatment of Epilepsy," *Psychiatric Quarterly* 2, no. 3 (1928): 307–13.

23. For instance: L.O. patient record, 1937: struck by car; K.N. patient record, 1937: hit with a baseball and bat; Z.S. patient record, 1940: gunshot wound, MNI.

24. The vast majority of Penfield's patients were Canadian and American. Fewer came from other countries.

25. For instance, B.L. patient record, 1936. B.L. was a patient at Presbyterian Hospital from 1918 to 1936 before arriving in Montreal.

26. L.O. patient record; V.G. patient record, 1931.

27. 1934 letter, in A.J. patient record, 1937.

28. During the Depression, beds were further subsidized by the city of Montreal and the province of Quebec and cost patients $3 a day. See A.J. patient record, 1937; "More Patients Cared for by Institute: Neurological Work Taxes Capacity of Hospital," *Montreal Daily Star*, Sept. 30, 1939, 3–4; "MNI Finances, 1938–9," folder E/PN 2-2 (1938–1949), Wilder Penfield Fonds, Osler Library for the History of Medicine (OLHM), McGill University, Montreal, QC.

29. L.O. patient record, 1937.

30. K.N. patient record, 1937.

31. Patient in this case was returned to a dehydration diet as part of a study at Harvard University. A.T. patient record, 1936.

32. On ketogenic and starvation diets: Eli Goldensohn et al., "The American Epilepsy Society: An Historic Perspective on 50 Years of Advances in Research." *Epilepsia* 38, no. 1 (1997): 124–50.

33. V.G. patient record, 1931.

34. The operative mortality rate for epilepsy surgeries between 1929 and 1939 was 4.2 percent. Other risks included acquired impairments and a worsening of the patient's epilepsy. Wilder Penfield et al., *Epilepsy and the Functional Anatomy of the Human Brain* (Boston: Little, Brown, 1954), 787–89.

35. J.V. patient record, 1936.

36. L.T. patient record, 1952.

37. L.T. patient record, 1952.

38. E.L. patient record, 1933.

39. E.L. patient record, 1933.

40. 1954–59 correspondence between Stanley Cobb and Wilder Penfield, Stanley Cobb Papers, HMS c53, series III subject files, box 30, folder 612, Countway Library for the History of Medicine (CLM), Harvard University, Boston.

41. "Cerebral Stimulation Studies Probe Memory," *Medical News*, June 23, 1958, 8–9.

42. Malcolm Macmillan, "A Wonderful Journey through Skull and Brains: The Travels of Mr. Gage's Tamping Iron," *Brain and Cognition* 5 (1986): 67–107.

43. See letters and telegrams between Wilder Penfield and his mother discussing Ruth's condition, 1919–1928: file D-C/D 29–15, Penfield Fonds, OLHM. See also Wilder Penfield et al., "Human Behavior after Extensive Bilateral Removal from the Frontal Lobes," *Archives of Neurology and Psychiatry* 44 (1940): 421–38.

44. William Feindel et al., "Epilepsy Surgery: Historical Highlights, 1900–2009," *Epilepsia* 50, suppl. 3 (2009): 131–51, 135–38.

45. Gere, " 'Nature's Experiment.' "

46. Wilder Penfield, "Epileptic Automatism and the Centrencephalic Integrating System," *Proceedings of the Association for Research in Nerve and Mental Diseases*, Dec. 1950, New York City, 513–28.

47. On John Hughlings Jackson: Owsei Temkin, *The Falling Sickness: A History of Epilepsy from the Greeks to the Beginnings of Modern Neurology* (Baltimore: Johns Hopkins University Press, 1945), 288–315.

48. Michael Bliss, *Harvey Cushing: A Life in Surgery* (New York: Oxford University Press, 2005).

49. On surgical techniques, dropping mortality rates, and the opening of the MNI: "McGill: Rockefeller Million Furthers Neurological Study," *Newsweek*, Aug. 18, 1934. Clipping in Folder E/PN 2.1, Wilder Penfield Fonds, OLHM.

50. Wilder Penfield, "Pitfalls and Success in Surgical Treatment of Focal Epilepsy," *British Medical Journal* 1, no. 5072 (1958): 669–72. On the importance of Foerster and Spanish technique: Penfield, *No Man Alone*, 156–58.

51. Penfield, *No Man Alone*, 156–58.

52. Wilder Penfield, "Psychical Seizures and Psychoparetic Automatism," speech at the meeting of the American Psychiatric Association, Montreal, May 23, 1949.

53. Penfield et al., *Epilepsy and the Functional Anatomy*; Peter Snyder and Henry Whitaker, "Neurologic Heuristics and Artistic Whimsy: The Cerebral Cartography of Wilder Penfield," *Journal of the History of Neurosciences* 22, no. 3 (2013): 277–91.

54. Winter, *Memory: Fragments of a Modern History*, 77.

55. N.N patient record,1951; L.T. patient record, 1952; M.A. patient record, 1951.

56. On the ancient religious and spiritual dimensions of epilepsy: Temkin, *Falling Sickness*.

57. J.V. patient record, 1936.

58. Penfield, "Psychical Seizures and Psychoparetic Automatism."

59. Wilder Penfield and Theodore Erikson, *Epilepsy and Cerebral Localization: A Study of the Mechanism, Treatment, and Prevention of Epileptic Seizures* (London: Bailliere, Tindall & Cox, 1941).

60. Wilder Penfield, "Memory Mechanisms," *Archives of Neurology and Psychiatry* 67 (Feb. 1952): 178–91.

61. Letter from Dr. Phanor Perot to Stanley Cobb, Apr. 14, 1960, box 30, folder 15, Correspondence with Wilder Penfield, 1960–67, Stanley Cobb Papers, CLM.

62. Correspondence between Penfield and a New York physician regarding aphasic patient and the study of speech, A.T. patient record, 1936.

63. Wilder Penfield, "Physiological Observations on the Cerebral Cortex of Man," Lane Medical Lectures, 29th Course, delivered Nov. 11–13, 17–18, 1947, Stanford University School of Medicine, San Francisco, CA; see file W/US, Wilder Penfield Archives, OLHM. For more on Penfield's operating theater and architecture: Annmarie Adams, "Designing Penfield: Inside the Montreal Neurological Institute," *Bulletin of the History of Medicine* 93, no. 2 (2019): 207–40.

64. On the parallel use of EEG in sleep and dream studies: Kenton Kroker, *The Sleep of Others and the Transformation of Sleep Research* (Toronto: University of Toronto Press, 2007).

65. V.G. patient record, 1931. Speaking of her case in an Aug. 24, 1956, letter, Penfield wrote, "Hers was one of the very first operations in which we produced a buzzing sound by stimulating the cortex, in what we came later to realize was the auditory area."

66. "Cerebral Stimulation Studies Probe Memory," *Medical News*, June 23, 1958, 8–9.

67. "Cerebral Stimulation Studies Probe Memory."

68. Penfield, "Physiological Observations on the Cerebral Cortex of Man."

69. On patient reliability in histories of psychiatry and psychology: Elizabeth Lunbeck, *The Psychiatric Persuasion: Knowledge, Gender, and Power in Modern America* (Princeton: Princeton University Press, 1995).

70. J.V. patient record, 1936.

71. A.T. patient record, 1936.

72. 1958 letter from Dr. Theodore Rasmussen to patient(s), D.B. patient record, 1951.

73. On patients' involvement and adeptness in neurological discourse: Stephen Jacyna and Stephen Casper, "The Psychasthenic Poet: Robert Nichols and His Neurologists," in Jacyna and Casper, *The Neurological Patient in History*.

74. B.K. patient record, 1936.

75. B.K. patient record, 1936.

76. This case is also detailed in Winter, *Memory: Fragments of a Modern History*, 78–79.

77. Letter from patient [initials redacted by author] to Penfield, June 23, 1949, [initials redacted] patient record, 1949.

78. The newspaper article cited here was enclosed as a clipping in the patient's June 23, 1949, letter to Penfield: W. Blakeslee, "Bared Under Anesthesia: Nerves Store Memory, Psychiatrists Report," (newspaper name not visible or mentioned by patient), May 25, 1949.

79. Letter from Penfield to patient [initials redacted by author] in 1950, [initials redacted] patient record, 1949.

80. On patient resistance to neurological standardization: Stephen Casper, "The Patient's Pitch: The Neurologists, the Tuning Fork, and Textbook Knowledge," in Jacyna and Casper, *The Neurological Patient in History*.

81. For a published version of Penfield's 1957 presentation, see Wilder Penfield, "Some Mechanisms of Consciousness Discovered During Electrical Stimulation of the Brain," *Proceedings of the National Academy of Sciences* 44, no. 2 (1958): 51–66. Before the presentation, Penfield asked J.V. to recount for the audience the event that had caused her fear as a girl and how it connected to her seizures. An excerpt of her response can be found in a 1972 report in her medical record following a visit to the MNI. See Nov. 14, 1972, report, J.V. patient record, 1936.

82. As recounted in Nov. 14, 1972, report, J.V. patient record, 1936.

83. Dwyer, "Stories of Epilepsy, 1880–1930"; Douglas C. Baynton, "Defectives in the Land: Disability and American Immigration Policy, 1882–1924," *Journal of American Ethnic History* 24, no. 3 (2005): 31–44.

84. Penfield, *No Man Alone,* 179–81, 228–30.

85. J.V. patient record.

86. 1931 letter, V.G. patient record, 1931.

87. Wilder Penfield, "Epilepsy and the Cerebral Lesions of Birth and Infancy" *Canadian Medical Association Journal* 41, no. 6 (1939): 527–34, 527.

88. 1958 letter to Penfield, A.L. patient record, 1958.

89. F.G.(2—second patient with these initials), patient record, 1942.

90. F.G.(2) patient record, 1942.

91. L.O. patient record, 1937.

92. Schlich, "Technological Fix."

93. Milton Silverman, "Now They're Exploring the Brain," *Saturday Evening Post,* Oct. 8, 1949.

94. "A Chance for Elizabeth," *Time*, Nov. 29, 1948, 73.

95. For the letters of those who wrote to Dr. Penfield around the time of the 1948 "Chance for Elizabeth," article in *Time*, see box 44, folder C/G 7, Penfield Fonds, OLHM. I have identified the letter writers cited in this chapter by single initials to protect their identities. Letter from letter writer L (of Washington, DC) to Penfield, Dec. 1948, box 44 C/G 7.

96. Letter from letter writer E (of London, ON) to Penfield, Dec. 1948, box 44, folder C/G 7.

97. A.G. patient record, 1936; M.S. patient record.

98. V.N. pathological report, *Neurosurgical Reports*, MNI, vol. 51, 1946.

99. 1938 letter to Penfield, K.M. patient record, 1937; August 1940 letter to Penfield; F.G.(3—third patient with these altered initials), patient record, 1942.

100. A.T. patient record, 1936.

101. December 1938 letter to Penfield, K.N. patient record, 1937.

102. December 1938 letter to Penfield, K.N. patient record, 1937.

103. Letter from Penfield, 1949, F.G.(1) patient record, 1949; V.G. patient record, 1931.

104. Vermont patient [initials redacted by author], patient record, 1936.

105. Vermont patient [initials redacted], patient record, 1936.

106. Vermont patient [initials redacted], patient record, 1936.

107. 1948 letter from Penfield, L.T. patient record, 1952.

108. For instance, one American patient was detained at St. Armand in 1935 and prohibited from entering Canada until Penfield intervened because he was "a prohibited person under subsection 'a' of section 3, as suffering from epilepsy." See June 1935 letters, B.E. patient record, 1936; F.G.(2) patient record, 1942.

109. O.P. patient record, 1945.

110. Letter from Penfield to J.V.'s father, Jan. 6, 1947, J.V. patient record, 1936.

111. J.B. patient record, 1946.

112. L.T. patient record, 1952.

113. Patient: "I still get the funny feelings, but no convulsions."

114. A.E. patient record, 1951; U.L. patient record, 1936.

115. Wilder Penfield and Kenneth W.E. Paine, "The Results of Surgical Therapy for Focal Epileptic Seizures," *Canadian Medical Association Journal* 73, no. 7 (1955): 515–31.

116. G.K. patient record, n.d.

117. E.L. patient record, 1933.

118. Z.N. patient record, 1936.

119. L.O. patient record, 1936.

120. On memory loss: J.B. patient record, 1946; on general disability: Q.K. patient record, n.d.

121. On suicide: L.S. patient record, 1950s; on depression: A.U. patient record, 1958; on institutionalization until 1958 death: A.L. patient record, 1935; on psychosis: H.E. patient record, 1935, L.O.(2) patient record, 1937.

122. E.H. patient record, 1934; L.O. patient record, 1936; E.L. patient record, 1933 (d. 1943).

123. Vermont patient, [initials redacted] patient record, 1936.

124. Vermont patient, [initials redacted] patient record, 1936.

125. Vermont patient, [initials redacted] patient record, 1936.

126. As cited in Lewis, *Something Hidden*.

127. Penfield, "Physiological Observations on the Cerebral Cortex of Man."

128. Wilder Penfield, "Observations on the Anatomy of Memory," summary note of address to the Montreal Medico-Chirurgical Society [draft], Nov. 4, 1949, W/U 113, Wilder Penfield Fonds, OLHM.

Epilogue

1. I have had several conversations over the years with my sister about her epilepsy. Although she often talks about landmark seizures—ones that happened on the first day of school or those following her first brain surgery—she tends to emphasize the broader constraints of living with epilepsy.

2. I refer here to examples from throughout this book, including Dr. Penfield's patients, people who sought driver's licenses, World War II veterans, and children at the White Special School. See also Joseph W. Schneider and Peter Conrad, *Having Epilepsy: The Experience of Control and Illness* (Philadelphia: Temple University Press, 1983), 104–41.

3. Lennard J. Davis, *Enforcing Normalcy: Disability, Deafness, and the Body* (New York:

Verso, 1995), xii–xvi; Susan Wendell, *The Rejected Body: Feminist Philosophical Reflections on Disability* (New York: Routledge, 1996); Emily K. Abel, *Sick and Tired: An Intimate History of Fatigue* (Chapel Hill: North Carolina University Press, 2021).

4. Christine Kenneally, "Do Brain Implants Change Your Identity?" *New Yorker*, Apr. 26 and May 3, 2021.

5. Kenneally, "Do Brain Implants Change Your Identity?"

6. For example, Milton Silverman, "We Can Lick Epilepsy," *Saturday Evening Post*, Jan. 17, 1946, 109–10.

7. H. D. Fabing and R. L. Barrow, *Epilepsy and the Law: A Proposal for Legal Reform in the Light of Medical Progress* (New York: Hoeber-Harper, 1956), 974.

8. Herbert Yahraes, *Epilepsy—The Ghost Is out of the Closet*, Public Affairs Pamphlet no. 98 (New York: Public Affairs Committee, 1944), 20.

9. Steven M. Spencer, "No Wonder Epileptics Are Bitter," *Saturday Evening Post*, Mar. 28, 1953.

10. Lewis P. Rowland, *The Legacy of Tracy J. Putnam and H. Houston Merritt: Modern Neurology in the United States* (New York: Oxford University Press, 2008).

11. Yahraes, *Epilepsy—The Ghost Is out of the Closet,* 20.

12. Yahraes, *Epilepsy—The Ghost Is out of the Closet,* 20.

13. For example, Tracy J. Putnam, "What the Patient Can Do to Help," chap. 3 in *Convulsive Seizures: How to Deal with Them* (Philadelphia: Lippincott, 1943), 55–61.

14. Yahraes, *Epilepsy—The Ghost Is out of the Closet,* 20.

15. Yahraes, *Epilepsy—The Ghost Is out of the Closet,* 20–21; *Annual Report of the White Special School, 1949–1950,* 82–83, Detroit Public School Records, Acc #681, Walter P. Reuther Library (WPR), Wayne State University, Detroit, MI.

16. Yahraes, *Epilepsy—The Ghost Is out of the Closet,* 21.

17. A.T. patient record, 1936, Montreal Neurological Institute (MNI), McGill University, Montreal, QC.

18. William G. Lennox, "Epilepsy: The Hopeful Disorder," *Parents' Magazine,* Jan. 1946, 124–27; Hans Hoff and Helen Clarke, *Will My Child Have Epilepsy?* (National Association to Control Epilepsy, April 1949), 1, New York Academy of Medicine Library (NYAM), New York.

19. Yahraes, *Epilepsy—The Ghost Is out of the Closet,* 7–8, 20.

20. Oral history interview with Isabel H. Baumann, 1980, State Archives, Wisconsin Historical Society Library (WHSL), University of Wisconsin, Madison.

21. Yahraes, *Epilepsy—The Ghost Is out of the Closet,* 21.

22. Yahraes, *Epilepsy—The Ghost Is out of the Closet,* 21.

23. Lennox, *Epilepsy and Related Disorders* (Boston: Little, Brown, 1960), 947.

24. Lennox, *Epilepsy and Related Disorders,* 928.

25. For example, Tridione was used at the White Special School between 1945 and 1947 before it was deemed unsafe and discontinued. Dilantin in some cases also caused hypoplasia of the gums or hirsutism. *Annual Report of the White Special School, 1946–1947,* 102, WPR.

26. Yahraes, *Epilepsy—The Ghost Is out of the Closet,* 21.

27. Lennox, "Epilepsy: The Hopeful Disorder," 125.

28. For example: "Some Plain English on Epilepsy," *Science Digest* 17 (1945): 82–83.

29. Yahraes, *Epilepsy—The Ghost Is out of the Closet*, 21.

30. Yahraes, *Epilepsy—The Ghost Is out of the Closet*, 21.

31. Yahraes, *Epilepsy—The Ghost Is out of the Closet*, 21.

32. Gerald Walker, "Epilepsy: Medical Triumph, Social Tragedy," *Pageant* 16, no. 6 (Dec. 1960): 90.

33. For example: William Lennox, "What the Minister Should Know About Epilepsy," *Pastoral Psychology* 44 (May 1954): 591–601.

34. Lennox, *Epilepsy and Related Disorders,* 957–62.

35. For example: Spencer, "No Wonder Epileptics Are Bitter"; Penfield Patient Records, MNI.

36. Putnam, *Convulsive Seizures,* 24.

37. On disability passing and the risks of exposure: Jeffrey A. Brune and Daniel J. Wilson, eds., *Disability and Passing: Blurring the Lines of Identity* (Philadelphia: Temple University Press, 2013).

38. Wilder Penfield, "Epilepsy and the Cerebral Lesions of Birth and Infancy," *Canadian Medical Association Journal* 41, no. 6 (1939): 527–34, 527; Edward M. Bridge, *Epilepsy and Convulsive Disorders in Children* (New York: McGraw-Hill, 1949); J.V. patient record, 1936, MNI.

39. Lennox, *Epilepsy and Related Disorders*, 928.

40. *Annual Report of the White Special School, 1955–1956*, WPR.

41. "Hearings of Epileptic Drivers, 1949–1954," Wisconsin Division of Motor Vehicles: Administrator's General Correspondence, 1949–1979, series 2328, WSHL.

42. W. G. Lennox et al., "Inheritance of Cerebral Dysrhythmia and Epilepsy," *Archives of Neurology and Psychiatry* 44, no. 6 (Dec. 1940): 1155–83,1181.

43. On disability activism in particular: Joseph P. Shapiro, *No Pity: People with Disabilities Forging a New Civil Rights Movement* (New York: Random House, 1994); Allison C. Carey, *On the Margins of Citizenship: Intellectual Disability and Civil Rights in Twentieth-Century America* (Philadelphia: Temple University Press, 2009); Margaret A. Winzer, *From Integration to Inclusion: A History of Special Education in the 20th Century* (Washington, DC: Gallaudet University Press, 2009).

44. Winzer, *From Integration to Inclusion*; "Epileptics to Meet, Map New Campaign," *Milwaukee Sentinel*, Sept. 17, 1954; "Fight Waged for Epileptics: They Aren't Mentally Ill, Milwaukee Doctor Tells Legislators" *Milwaukee Journal*, May 5, 1955.

45. John Paul Leach and Rebecca O'Dwyer, *Epilepsy Simplified* (Malta: Gutenberg Press, 2011), 56–114; Orrin Devinksy, *A Guide to Understanding and Living with Epilepsy* (Philadelphia: F.A. Davis, 1994); John Freeman et al., *Seizures and Epilepsy in Childhood: A Guide for Parents* (Baltimore: Johns Hopkins University Press, 1997).

46. For examples of educational efforts countering the myths surrounding epilepsy: Joseph I. Sirven and Patty Obsorne Shafer, "Facts and Statistics About Epilepsy," Epilepsy Foundation, July 6, 2013, revised by Andres M. Kanner, Elaine Wirrell, and Patty Obsorne Shafer, Feb. 27, 2019, https://www.epilepsy.com/what-is-epilepsy/statistics; Leach and O'Dwyer, *Epilepsy Simplified*, 1; Devinsky, *A Guide to Understanding and Living with Epilepsy*, 4–6.

47. For instance: letter from Mrs. James Hanson to Mr. Lowell Jackson, July 7, 1979, Wisconsin Department of Transportation Records, WHSL; Kenneally, "Do Brain Implants Change Your Identity?"

48. Lennox, *Epilepsy and Related Disorders*, 958.

49. Robert Aronowitz, "The Converged Experience of Risk and Disease," *Milbank Quarterly* 87, no. 2 (2009): 417–42; Thomas Schlich and Ulrich Tohler, *The Risks of Medical Innovation: Risk Perception and Assessment in Historical Contexts* (New York: Routledge, 2005).

Index

Page numbers in *italic* refer to figures.

absence seizures, vii, 168n15
"AC 186," 92
African American children: racial segregation of, 77–78, 95
Alder, Ken, 63, 188n117
A. L. Holmes School, 196n102
American Association of Motor Vehicle Administrators (AAMVA), 109, 110
American Epilepsy League (AEL), 32, 35, 58
American Epilepsy Society (AES), 35
American Psychiatric Association, 139, 145
Americans with Disabilities Act (ADA), 164, 165
anticonvulsant medications: development of, 10, 30–31, 52, 176n58; dispensation of, 98; popularity of, 159; seizure control effect of, 3, 31, 39, 75–76, 159–60, 168n12; testing of, 80–81. *See also individual drugs*
Are Seizures a Safety Hazard? report, 88–89
Argosy (magazine), 21, 173n10
Atlas of Electroencephalography, 48, 49
autism, 84
automobiles: accidents and fatalities related to, 103, 104, 108, 111, 115; cultural histories of, 102; growth of ownership, 103; safety features, 104
Ayres, Lew, *23*

barbiturates, 31, 80
Barrow, Roscoe, 120, 122; *Epilepsy and the Law*, 119
Barrymore, Lionel, 22, *23*
Berger, Dorothy, 59, 186n79
Berger, Hans, 50, 57, 59, 66
Borck, Cornelius, 182n12
Boster, Dea, 10, 172n67
Boston Children's Hospital, 186n85
brain: complexity of, 137; electrical activity of, 1, 4; experiments on, 137, 138; stimulations of, 137; studies of, 130, 139–40, 142, *143*; tissue removed during neurosurgery, 134

brain waves: clinically "abnormal," 65; dysrhythmic, 15, 61; EEG studies of, 30, 47, *56, 57*, 58, 62, 65, 71, 72; of geniuses, 68, 69; interpretation of, 49, 50, 51, 59, 60; "normal," 64
Brand, Max: *Dr. Kildare's Crisis*, 19, 41, 56–57, 64, 173n10
breathalyzer tests, 104
Brown v. Board of Education, 192n13
Buck, Carrie, 26
Buck v. Bell, 26, 28

Caesar, Julius, 37
California: Bay Bridge accident in, 99; compulsory reporting of epilepsy in, 99, 102, 105, 106, 108, 109; driver licensing laws, 15, 99–100, 102–3, 105; sterilization law, 26
Canada: disability rights in, ix; eugenic sterilization laws, 146; refusal of entry for people with epilepsy, 152, 211n108
Carter, W. E., 102, 103, 104, 105, 106, 107
cerebral cortex: electrical stimulation of, 132, 138–40; mapping of, 137, 139; studies of, 130, 142
cerebral dysrhythmia: diagnosis of, 71; disclosure of, 164; EEG's reading of, 30, 55, 58, 64; manifestation of, 57; notions of heredity, 59, 60, 66
"Chance for Elizabeth, A" (*Time* magazine article), 149
Charcot, Jean-Martin, 67
children: EEG testing of, 59; protection from seeing seizures, 41, 78
chronic fatigue syndrome, 12
chronic illness, xi, 9, 12, 44, 165
civilian epileptic, 87
civil rights movements, 13
Clemens, Jean, 26–27, 29
Cobb, Stanley, 134, 137, 140, 177n70
Codarre, Edwin: trial of, 62–63